CHATGPT TO DOUBLE YOUR BUSINESS IN 90 DAYS

THE HANDS-ON AI AND AUTOMATION GUIDE TO EXPLOSIVE GROWTH IN TODAY'S ECONOMY

ETHAN KING

International Bestselling Author of *Wealth Beyond Money*

PRE-RELEASE EDITION

CHATGPT TO DOUBLE YOUR BUSINESS IN 90 DAYS

Copyright © 2024 by Ethan King

SIMPLE Success Systems LLC
Atlanta, GA

All rights reserved. No part of this book may be reproduced or transmitted in any form or by any means without written permission from the author.

ISBN 978-0-9856056-6-7 (paperback)

ISBN 978-0-9856056-7-4 (hardcover)

What People Are Saying

"...it's like a cheat code for growth!"

—Trent Clark, 3x World Series Coach, CEO & Bloom Growth Coach at Leadershipity, TEDx and Keynote Speaker, Host of Winners Find A Way podcast, and Author of Leading Winning Teams (Wiley)

"Ethan King is a great leader who helps foster other leaders. His new book is an important one to have in the hands of every entrepreneur for the modern era."

—Moe Rock CEO of the Los Angeles Tribune

"...a game-changer for anyone looking to harness the power of AI to elevate their business..."

—Jami Lah, Executive Producer, TEDxStGeorge

"...a bold playbook for leaders to harness AI, optimize work, and drive exponential growth. Learn to leverage AI or get left behind."

—Gene Hammett, Keynote Speaker, Executive Coach to Growth Companies, and Co-author of *How to Have Tough Conversations*

"...a must-read for anyone looking to elevate their business communication."

—Ryan Bean, TEDx Speaker, Meditation & Breathwork Instructor on Insight Timer, YouTube, and the Source app

"...packed with actionable insights and innovative strategies, this book is an essential read for anyone looking to harness AI to transform their business."

—Melanie C Graf, High Performance Coach and Speaker
https://www.melaniegraf.com/

"An invaluable toolkit packed with clear, actionable steps perfectly aligned with the four keys to business growth: lead generation, conversion, customer ascension, and retention..."

—Celeste Jonson, International Speaker, Author of *D.A.R.E. to Succeed Despite the Odds*, Leadership Presence Coach, and HOPE Ambassador

"...a comprehensive, straightforward path to driving real efficiency and exponential growth in your business."

—Parth Patel, CEO, Six Consulting, Inc.

TABLE OF CONTENTS

- Preface 1
- Get Over-the-Air Updates for This Book 4
- The Power of AI and Automation 5
  - How AI is Transforming Industries and Our Lives 6
  - Iron Man's Jarvis Becomes Reality 6
  - Automation in Scaling Businesses 7
  - Key Takeaways 9
- How To Get The Most Out Of This Book 10
- Understanding ChatGPT and LLMs 12
  - What is ChatGPT? 12
  - ChatGPT Alternatives 14
  - The Importance of Understanding AI Philosophy and Communication ... 15
  - Key Takeaways 16
- Is It Ethical To Use AI? 17
- Train Your AI To Be YOU (Do This First!) 19
  - Why Clone Yourself? 19
  - Setting Up Custom Instructions 20
- Double Your Business: Just One More _____ 29
- Prompt Engineering 101 36
- Tools To Increase Number of Leads 39
  - Day 1: HARO Hack—Get Free Media Coverage and Content 40
  - Day 2: Press Releases 42
  - Day 3: Blog Posts 43
  - Day 4: Email Newsletters 44
  - Day 5: LinkedIn Articles 45
  - Day 6: Medium Articles 46
  - Day 7: Get Featured On Influential News Sites 48
  - Day 8: Answer Questions on Quora 49
  - Day 9: X/Tweets 51
  - Day 10: Instagram Threads 52
  - Day 11: Batch Create Text-Image Posts In Seconds 53
  - Day 12: Live Videos 55
  - Day 13: Long-form Videos 57
  - Day 14: Automated Webinars 58
  - Day 15: Podcasting The Easy Way 62
  - Day 16: Create Short-Form Videos With Less Effort 65
  - Day 17: Publish Short-form Videos On Autopilot 68
  - Day 18: Post to Social Media Stories Daily 69

- Day 19: Viral TikTok Videos...72
- Day 20: Be A First-Mover On New Social Media Apps..........................73
- Day 21: Respond to Comments and Reviews.....................................73
- Day 22: Deploy AI Chatbots for Website Engagement.........................75
- Day 23: AI-Driven SEO Optimization..77
- Day 24: Ask Your Existing Clients For Referrals..................................80
- Day 25: Newsjacking...83
- Day 26: Be A Guest On Podcasts...86
- Day 27: Trade Shows..88
- Day 28: Write Case Studies...92
- Day 29: Buy Leads..94
- Day 30: Cold Calling Scripts..96
- Day 31: Billboards..98
- Day 32: Direct Mail..100
- Day 33: Influencer Marketing...103
- Day 34: Boost Organic Posts Into Paid Ads.....................................106
- Day 35: Write a Book..108
- Day 36: Creating Images With AI...110
- Day 37: Design Your Book Cover..116
- Day 38: Create Logos From Scratch With AI...................................119
- Day 39: Use Your Sketches To Create Logos...................................123
- Day 40: Create A Website In Seconds, Simply By Describing It.......124

Tools To Convert Leads Into Paying Clients...125
- **Day 41: Master Storytelling Science..126**
- Day 42: Email Drip Campaigns..127
- Day 43: Followup.cc..130
- Day 44: Direct Messaging..134
- Day 45: Promotional Products...137
- Day 46: Virtual and In-Person Events..140
- Day 47: Personalized SMS Campaigns...143
- Day 48: Retargeting Ads Turn Browsers into Buyers........................146
- Day 49: Voicebots...149
- Day 50: Automated Survey Follow-Ups...152
- Day 51: Copywriting for Website Pages..155
- Day 52: Personalized AI Chatbots..158
- Day 53: Create Proposals..160
- Day 54: Automated Sales Workflows..164
- Day 55: Abandoned Cart Emails..165
- Day 56: Motivating Your Sales Team..168
- Day 57: Competitive Analysis..171
- Day 58: New Product Research & Development.............................172
- Day 59: Make A Pitch Deck To Raise Capital..................................174
                                                                                                          175

Tools to Increase Average Order Value 177
- Day 60: Dynamic Pricing Models 178
- Day 61: Premium Service Tiers 179
- Day 62: Pay-In-Full Discounts 180
- Day 63: Skip-the-Line Fees 181
- Day 64: VIP Service Levels 182
- Day 65: Product Bundling 183
- Day 66: Extended Warranties 185
- Day 67: Guarantees 187
- Day 68: Order Bumps 189
- Day 69: High-Ticket Upsells 192
- Day 70: AI-Generated Product Descriptions 194
- Day 71: Automated Customer Feedback 196

Tools to Increase Repeat Purchase Frequency and Retention 197
- Day 72: Interactive and Gamified Content 198
- Day 73: Apply for Grants 201
- Day 74: Special Offers and Coupons 204
- Day 75: Subscriptions 205
- Day 76: Regular Learning Content 207
- Day 77: 100-Day Onboarding Plan 208
- Day 78: After-Purchase Product Suggestion Emails 211
- Day 79: AI-Driven Loyalty Programs 212
- Day 80: Hiring Right 214
- Day 81: Leadership 216
- Day 82: Core Values 218
- Day 83: Performance Reviews 220
- Day 84: Handwritten Thank You Notes 222
- Day 85: Birthday and Anniversary Offers 224
- Day 86: Feedback and Improvement Surveys 225
- Day 87: Build An Online Community 228
- Day 88: Re-Engagement Campaigns 232
- Day 89: Personalized Video Messages 234
- Day 90: Endless AI tools 235

Create Your Own Custom GPTs 236
- Getting Started With GPTs 237
- Sharing and Monetizing Your GPTs 237
- How to build your own custom ChatGPT 238
- Examples of Custom GPTs 238

Creating Your Business Advisor GPT 240
- Exercise: Informing the Business Growth Strategist GPT 241
- Leveraging the Business Growth Strategist GPT 242

The Road Ahead: Future Predictions 243

    Integration Into Daily Life..243
    Advanced AI Interactions..243
    Creative Capabilities..243
    Human Interaction and Trust..244
    Enhanced Search and Personalization..244
    Virtual and Augmented Reality...244
    Competition Among AI Providers...245
    Predictive and Proactive AI..245
    Ethical and Societal Implications...246
    Future Business Applications...246
    Conclusion..247

Designing Your Future Self..247
    The Power of Decisions..**248**
    Step 1: Describe Your Evolved Self..248
    Step 2: Create Your "Future You" GPT...250
    Step 3: Consult Your "Future You" GPT..251
    You Can Have It All..251

About the Author...252
Resources..**253**
    SimpleSuccess.ai..**254**
    EthanKing.com...254
    Your Evolve Shirt...254
    SIMPLE Success School...254
    Over-the-Air Updates for This Book..255
Reviews..255
    ....**256**

Preface

Why this book? Why now?

In the early 2010's (way before ChatGPT), I was struggling to grow our little embroidery business. We had about 13 employees, a hefty mortgage for the new office building we had just purchased, and were only making about $400,000 in revenue. We were overstaffed and we weren't making any profits. In fact, we were losing money each year and stacking on debt, in hopes that something would change.

Well something did change, but not in a good way.

Out of nowhere, we were slapped with an unfair lawsuit that forced us to stack legal expenses on top of our already-struggling profit and loss statement. Cash flow and profits went deeper into the negative, while debt increased. At one point, an advisor even hinted we should consider filing bankruptcy.

Between letting some staff go and simply not hiring replacements when other employees left, we went down to just six team members. And a lot of the day-to-day work ended up back on my shoulders, which meant 12-16 hour work days for me, often without paychecks for myself or my wife, because we always put our team members first. I've learned that many small business operators end up in a similar situation at some point. Does this sound familiar to you?

Forced to do more with less, I dove headfirst into the world of marketing automation. Little did I know how critical these skills would become.

A few years later, through sheer determination and the strategic use of automation tools, I managed to double my business's revenue while cutting my staff in half. We crossed the $1,000,000 annual revenue mark with just six team members. This means our revenue per employee went from roughly $30,769 to $166,667 in the span of about 3 years. We achieved profitability and got out of debt. This also freed up our lifestyle a lot, and my wife and I both now drive our dream cars, travel the world with our kids, and are able to provide them with experiences that we couldn't afford before.

And not only is our embroidery business still thriving, we have also started more businesses that are currently growing at exponential rates.

This experience taught me that with the right tools and mindset, it's possible to not just survive, but thrive in the face of adversity. And if I can do it, so can you.

Here we are again at another peculiar time in our economy. At the time of this writing, inflation rates are up and everything costs more, wage demands are high, and consumer spending is low. This slow-brewing perfect storm has many small businesses struggling to survive, and unfortunately, many have been forced to close their doors.

According to the U.S. Bureau of Labor Statistics, business failure rates have been increasing and in 2023 are more than double what they were in 2019.

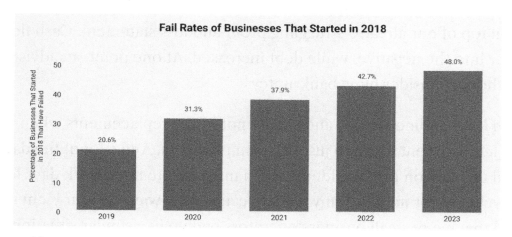

Source: bls.gov via https://www.shopify.com/blog/percentage-of-businesses-that-fail

The good news: we are now on the brink of another revolutionary shift in business operations with the advent of Large Language Models (LLMs) like ChatGPT. These powerful AI tools, combined with the latest advancements in automation, have the potential to transform the way businesses operate. I predict that in the near future, we will see a surge in multimillion-dollar businesses run by just one person and a virtual assistant (VA). This isn't a far-off fantasy—it's an emerging reality that you can tap into right now.

The tools are here. They are powerful, accessible, and designed to help you achieve remarkable results with minimal overhead. But here's the catch: you have to use them and use them correctly.

It reminds me of the old joke about the preacher in the flood. As the heavy rains poured, a bus came by and told him "get in!" The preacher refused and said "God will save me." Then as the water rose, a boat came by and told him "get in!" The preacher refused and said "God will save me." Then as the water rose up to his roof, a rescue helicopter came by, offering him a way out. The preacher again refused, saying "God will save me."

The preacher drowned.

When he asked God "why didn't you save me?" God replied, "What do you mean? I sent you a bus, a boat, and a helicopter!"

Don't be like that preacher. The tools are here; you just need to use them, and this book will teach you how to do it right.

I'm not perched on a mountaintop spewing theory; I'm in the trenches applying these tools to my businesses every day. The road of entrepreneurship is rough, and we need every advantage we can get to help us win.

I will guide you through the practical steps and strategies to harness the power of ChatGPT and other AI tools. You'll learn how to optimize your workflow, enhance customer engagement, and make data-driven decisions that propel your business forward.

Will you double your business in 90 days? I believe that if you fully apply and implement each strategy in this book, you can more than double your revenue. If you actually execute every step and your business still doesn't grow, then reach out and I will coach you personally to help you get there.

This isn't just about working harder; it's about working smarter. By leveraging these tools, you can achieve extraordinary growth with minimal effort and expense. This book is your roadmap to transforming your business and achieving the success you've always envisioned.

Let's get started.

Get Over-the-Air Updates for This Book

AI changes fast.

Like *really* fast.

Imagine if this book you're reading right now could update itself, just like a Tesla or smartphone does overnight. Guess what? It can!

We are constantly refreshing sections of this book to keep you ahead of the curve. To get these free "over-the-air" updates sent directly to your email inbox, just head over to **ChatGPTBookUpdates.com** or scan the QR code below.

This way, you will stay updated effortlessly as AI continues to transform our businesses and lives.

The Power of AI and Automation

Artificial Intelligence (AI) might seem like a modern marvel, but its roots actually stretch back decades. The concept of AI was first introduced in the 1950s by pioneers like Alan Turing and John McCarthy. Turing's famous question, "Can machines think?" set the stage for decades of exploration into machine learning and computational intelligence. Fast forward to the 21st century, and we've seen AI evolve from simple programs capable of playing chess to complex systems that can learn, adapt, and even drive cars.

I love the self-driving feature in my Tesla, and while it still has some ways to go, it constantly improves itself by learning from trillions of data points around the world.

AI is a buzzword right now, kind of like how "Internet" was the buzzword of the 1990s. In the early days of the internet. It was a novelty, something only a few had access to. But as technology advanced, it became an essential tool for communication, commerce, and entertainment. The internet is so woven into our everyday lives that we now take it for granted, like air or water.

Similarly, AI has now evolved from a catchy sci fi term to a practical, everyday tool. Today, AI powers everything from our smartphones' voice assistants to sophisticated data analysis in industries like finance and healthcare.

This rapid evolution of AI is a testament to human ingenuity and our relentless pursuit of innovation. It also highlights the potential of AI to continue to revolutionize the way we live and work, making it a critical component of any forward-thinking business strategy.

HOW AI IS TRANSFORMING INDUSTRIES AND OUR LIVES

AI's impact is visible across various industries, transforming processes and creating efficiencies that were once unimaginable.

In healthcare, AI is used to analyze medical images, predict patient outcomes, and even assist in surgeries. This technology not only improves accuracy but also saves lives by enabling early detection of diseases.

In the retail sector, AI helps businesses understand consumer behavior through data analysis, leading to personalized shopping experiences. Companies like Amazon and Netflix leverage AI to recommend products and content based on past behavior, significantly enhancing customer satisfaction and loyalty. Amazon's recommendation engine, for example, is responsible for 35% of its total sales[1].

AI is also making waves in our daily lives. From virtual assistants like Siri and Alexa that manage our schedules to smart home devices that learn our preferences and adjust settings accordingly, AI is seamlessly integrating into our routines. This pervasive influence of AI underscores its transformative potential and its ability to enhance both personal and professional aspects of our lives.

IRON MAN'S JARVIS BECOMES REALITY

My favorite superhero is Iron Man. When the first movie came out in 2008, I thought Tony Stark was so cool. I always wanted an AI assistant like Jarvis, and now we have it (or at least we are getting close).

Today, anyone can set up a voice assistant like Alexa, Siri, or Google to power your smart home. What was once science fiction is now normal.

---

[1] "The power of product recommendation and why your online business should implement it in 2022" *ExposeBox*, March 6, 2022, https://exposebox.com/the-power-of-product-recommendation-2022/

I've even changed the settings of my iPhone's Siri voice to have a male British accent so it sounds more like Jarvis. And I have it refer to me as "sir" (just say "Siri, call me Sir from now on" if you want to do this too). I know, nerd.

I also often have voice conversations with the ChatGPT app on my phone. Not just about business. I have it to help me think through travel plans, help diagnose body aches I might be having, act as my therapist, life coach, writing assistant, and so much more.

Imagine having one or several Jarvis-like assistants for your business. This AI wouldn't just manage mundane tasks but could also provide insights, predict trends, and even suggest strategic moves.

We will cover all of that in more in this book.

AUTOMATION IN SCALING BUSINESSES

In my book *Wealth Beyond Money*, I teach the E.A.T. method for freeing up your time. Examine every task you do, and first ask yourself "can I eliminate this?" If not, then automate it. If you can't automate it, then train someone else to do it.

- Eliminate it, or
- Automate it, or
- Train someone else to do it

You'd be surprised at what you can automate on your own, without knowing any programming code.

Automation is a must for businesses looking to scale. By automating repetitive tasks, businesses can free up valuable time and resources, allowing employees to focus on more strategic activities. This shift not only enhances productivity but also drives innovation and growth.

Klarna, the Swedish-based fintech company known for its "buy now, pay later" service, is leveraging generative AI to save millions annually on marketing costs. By automating campaigns and generating images with AI tools like Midjourney, DALL-E, and Firefly, Klarna cut its sales and marketing budget by 11% in the first quarter, with AI responsible for 37% of the savings.[2]

Klarna has also rolled out an AI Assistant that's revolutionizing its customer service operations. Since its launch, this tool has handled 2.3 million conversations, covering tasks like refunds, returns, payment issues, and more, across 35 languages, 24/7. This AI Assistant is effectively doing the work of 700 full-time customer service reps.

This technology has drastically reduced the time to resolve customer issues from 11 minutes to less than two minutes and cut repeat inquiries by 25%. Klarna's AI-driven approach is saving the company more than $40 million annually, a return on investment that showcases the transformative potential of AI.[3]

While small and mid-sized businesses might not afford such technology now, the increasing adoption by big brands will eventually make these AI tools more accessible and affordable. But you can actually achieve a lot of this now, with tools like ChatGPT, Zapier, Make.com, and others we will discuss in this book.

Automation extends beyond just marketing and customer service.

In manufacturing, AI-driven robots streamline production processes, ensuring consistency and reducing errors. For example, Tesla's Gigafactories are highly automated, with AI and robotics playing a crucial role in manufacturing electric vehicles. This automation allows Tesla to scale production rapidly while maintaining quality standards.

In finance, automated systems manage transactions, detect fraud, and provide financial insights, enabling businesses to operate more efficiently and securely. PayPal, for instance, uses AI to detect fraudulent transactions in real-time, reducing losses and enhancing the security of its platform.

---

[2] Supantha Mukherjee, "AI Helping Klarna Cut Marketing Costs by $10 Million," *Inc.com*, May 28, 2024, https://www.inc.com/reuters/ai-helping-klarna-cut-marketing-costs-by-10-million.html

[3] Gene Marks, "Klarna's New AI Tool Does The Work Of 700 Customer Service Reps," *Forbes.com*, March 13, 2024, https://www.forbes.com/sites/quickerbettertech/2024/03/13/klarnas-new-ai-tool-does-the-work-of-700-customer-service-reps/

These examples highlight the significant role of automation in scaling businesses. By embracing AI and automation, businesses can achieve greater efficiency, reduce costs, and position themselves for sustained growth in an increasingly competitive market.

KEY TAKEAWAYS

- Like the internet's rise in the '90s, AI has become essential, powering our smartphones, assisting in healthcare, and analyzing data in finance.
- AI transforms industries by improving efficiency and accuracy, such as analyzing medical images in healthcare or personalizing shopping experiences in retail.
- Tools like Siri, Alexa, and ChatGPT bring us closer to the AI capabilities of Iron Man's Jarvis, assisting in everything from mundane tasks to strategic decision-making.
- Using the E.A.T. method—Eliminate, Automate, Train—can free up time and resources, enhancing productivity and driving business growth.
- Case Study: Klarna uses AI to save millions annually, automating marketing and customer service tasks, showcasing AI's potential for cost savings and efficiency.
- While high-end AI tools are pricey, affordable options like ChatGPT, Zapier, and Make.com can help businesses achieve significant automation benefits now.
- From Tesla's automated manufacturing to PayPal's fraud detection, AI-driven automation is crucial for maintaining quality and efficiency.
- As big brands adopt AI, these tools will become more accessible, allowing even small businesses to benefit from AI-driven efficiencies and growth.

By understanding and harnessing the power of AI and automation, businesses can navigate the complexities of the modern market, drive innovation, and achieve exponential growth.

The future is here now. Don't get left behind.

How To Get The Most Out Of This Book

This is a workbook, not a fiction novel. It wasn't designed for you to read leisurely in one pass. Read a step, then stop, open your computer, and do the work.

This book is designed for you to follow the steps in order. Most of these strategies build on the preceding ones, ensuring a smoother and more effective implementation. However, depending on where you are in your AI journey, you might already be familiar with some of the steps.

To keep things manageable, I've structured it so you can tackle one step per day for 90 days. This allows you to really dive into each tactic and explore its possibilities. But if you're eager to move faster, you can certainly speed up the pace. I don't recommend moving slower, because things are evolving so fast that if you take longer than 90 days to implement or at least try these tactics, many of the processes will have already changed.

By the end of this journey, you'll have prompts and GPTs in place acting as your assistants, helping you get things done more efficiently. The initial groundwork might take some time, especially if you're new to this, but once you understand the foundations, everything will fall into place.

Don't be intimidated by AI. This book is not here to overwhelm you but to make your life easier. No programming knowledge is needed; if you can copy and paste, you can do everything I will show you.

I'm not a programmer myself. In fact, I was an Art major in college, got kicked out for a semester, arrested multiple times, and was unemployed for almost two years afterward, taking out the trash at a strip club to make ends meet. If I can do this, so can you.

I'm tool-agnostic; I'm just sharing what works best right now. The landscape changes fast, but the philosophy and principles in this book remain constant. Some people fear

AI will take their jobs; others embrace it. It's here to stay, so you might as well embrace it too.

Here's a story: I once hired a store manager with an impressive resume. She looked great on paper and interviewed well. One day, I had her do inventory. She showed me her spreadsheet. While looking over her shoulder, I asked her to give me the total number of a particular item. Instead of doing a quick SUM calculation, she started counting the cells by hand and adding up the numbers in her head. I couldn't believe it! Needless to say, I quickly let her go.

Today, I'd hire the person or firm that embraces AI over the one that doesn't.

Would you hire a CPA who counts your money by handwritten multiplication instead of using accounting software and spreadsheets?

The choice is clear: AI is not just a tool for efficiency, it's a necessity for staying competitive in today's market. Let's get started and transform the way you do business.

Understanding ChatGPT and LLMs

In today's fast-paced digital world, understanding the tools at our disposal can make the difference between thriving and merely surviving. One such tool, ChatGPT, has transformed how businesses operate, innovate, and communicate. Let's dive into what ChatGPT is, its key features, common misconceptions, and why understanding the philosophy of AI is crucial.

WHAT IS CHATGPT?

Imagine having a really smart, multilingual friend who has read all the books and absorbed the vast knowledge of the Internet (with the Plus version). This friend is always available, ready to help you with anything you need, from answering complex questions to assisting with daily tasks. That's ChatGPT in a nutshell.

GPT stands for Generative Pre-trained Transformer. It's a bit of a mouthful, but it breaks down like this:

- Generative: It can generate text based on the input it receives.
- Pre-trained: It has been trained on a vast amount of data before you even start using it, so it comes equipped with a broad base of knowledge.
- Transformer: This refers to the type of model architecture it uses, which is particularly good at understanding and generating human-like text.

You can ask ChatGPT almost anything, and it'll do its best to help. Whether you need assistance drafting an email, generating creative ideas, or even debugging code, ChatGPT is there for you. However, it's important to understand that ChatGPT has its quirks.

For one, it sometimes won't ask you clarifying questions unless you tell it to. This means it might take a stab at your request and potentially get things wrong. This phenomenon, where the model confidently provides incorrect information, is known as "hallucination." It's a reminder that while ChatGPT is powerful, it's not infallible.

Think of it this way: You know how when you start typing something into Google search, it generates a list of phrases, trying to predict what you're searching for? That's called predictive text. ChatGPT is like predictive text on steroids. It doesn't just finish your sentences; it can draft entire paragraphs, engage in conversations, and provide detailed explanations on a wide range of topics.

However, it's crucial to remember that ChatGPT is essentially a language calculator. It's a tool designed to process and generate text based on patterns it has learned. It doesn't have feelings, emotions, or personal experiences. It's not sentient; it doesn't think or understand in the way humans do.

One of the fascinating aspects of ChatGPT is its ability to role-play. It can think, act, and speak like any expert or famous person, living or dead. Want to get advice from Albert Einstein on your physics homework? Or perhaps you need some business insights from Steve Jobs? ChatGPT can simulate these personas, offering insights and perspectives as if you were conversing with the actual individuals.

Using ChatGPT effectively involves understanding its strengths and limitations. It's a remarkable tool for generating content, brainstorming ideas, and providing quick information. But it's also essential to critically evaluate its responses and cross-check information, especially for critical tasks.

With the right approach, ChatGPT can transform the way you work, freeing up your time to focus on strategic, high-impact activities.

You can get started immediately, for free, at https://chatgpt.com/

For just $20/month (a steal), the Plus version gets you more features like:

- Priority Access: Faster responses, even during busy times.
- Advanced Features: Access the latest and greatest AI models.
- Expanded Knowledge: More up-to-date information and data. It can search the web real time.

- Create and Use Custom GPTs: AI models tailored to your needs.
- Higher Limits: More requests and longer sessions.
- Better Security: Extra protection for your data.
- Multimodal: Generate and process images, audio, and video (if available).
- Regular Updates: New features and improvements regularly.

IMPORTANT: You will need the paid version of ChatGPT to carry out all of the steps in this book.

CHATGPT ALTERNATIVES

While ChatGPT is a powerful tool, exploring alternatives can help find the best fit for your business needs. Here are some of the most recent and noteworthy alternatives:

- **Google Gemini:** Formerly known as Bard, Google Gemini offers deep integration with Google services like Gmail and Google Docs. It excels in tasks such as drafting documents, analyzing data, and creating presentations. With real-time information access and the ability to double-check responses via Google search, it is a robust tool for busy professionals and remote workers.
- **Microsoft Copilot:** Integrated into Microsoft 365 apps, Copilot uses GPT-4 architecture to assist with tasks like drafting documents, analyzing data, and scheduling meetings. It also provides image generation and analysis capabilities, making it a versatile tool for various business applications.
- **Claude 3:** Developed by Anthropic, Claude 3 excels in natural and extended conversations, summarization, editing, and decision-making. It supports large context windows and non-English languages, making it suitable for international use.
- **Perplexity AI:** This AI tool is designed to excel in searching the web and providing accurate, contextually relevant information. It is particularly useful for businesses that require up-to-date information and quick answers to complex queries. Perplexity AI stands out for its ability to conduct real-time web searches and provide detailed responses, making it an excellent choice for research-intensive tasks.

- **Grok:** Grok is your AI-powered assistant designed to streamline and enhance your business operations. Owned and operated by Elon Musk's AI startup, xAI, Grok leverages advanced AI technologies like natural language processing and GPT to understand and generate human-like text, integrate data from various sources, and adapt to your specific business needs with customizable workflows. It's user-friendly, scalable, and secure, making it an indispensable tool for managing tasks, analyzing data, and improving efficiency. Grok continuously learns and evolves, ensuring it gets better over time, and seamlessly integrates with your existing systems, providing a comprehensive solution for modern business challenges.

There are other alternatives like Jasper and Writesonic, but many of them are powered by Open AI/ChatGPT anyway, so I won't list them here.

As of right now, ChatGPT is the dominant tool. It's like the swiss army knife of AI.

The other tools are good for certain things—like Claude 3 sometimes produces more human-sounding results, or Perplexity can be better for deep research—but ChatGPT can be customized to do all of those things well too.

THE IMPORTANCE OF UNDERSTANDING AI PHILOSOPHY AND COMMUNICATION

AI technology will continue to evolve, with new tools emerging over time. Thus, understanding the underlying philosophy and learning how to communicate with AI is vital.

Think of AI as a highly intelligent assistant. The clearer and more precise your instructions, the better the results. This is evident in how large corporations like Amazon use AI. By training their teams to interact effectively with AI systems, they ensure seamless operations and innovation.

Moreover, grasping AI philosophy helps in anticipating future trends and staying ahead of the curve. It encourages a mindset of continuous learning and adaptation, crucial for long-term success.

KEY TAKEAWAYS

- Understanding ChatGPT: Think of ChatGPT as a highly knowledgeable, multilingual friend who's always available to help, thanks to its Generative Pre-trained Transformer (GPT) technology.
- Capabilities and Limitations: ChatGPT can generate detailed text, engage in conversations, and provide expert advice, but it can sometimes produce incorrect information (hallucinations) and lacks human emotions and experiences.
- Role-Playing and Practical Uses: ChatGPT excels at role-playing, simulating experts' advice, and performing tasks like content generation and brainstorming. It's essential to evaluate its responses critically and cross-check information for accuracy.
- AI Alternatives and Their Strengths: While ChatGPT is a versatile tool, exploring alternatives like Google Gemini, Microsoft Copilot, Claude 3, Perplexity AI, and Grok can provide specialized solutions tailored to specific business needs. Understanding AI's philosophy and communication techniques is crucial for maximizing these tools' potential and staying competitive.

Remember, in the rapidly evolving digital landscape, mastering the use of AI tools like ChatGPT isn't just an advantage—it's a necessity for staying competitive and driving your business forward. Let's dive in and transform the way you work, one step at a time.

Is It Ethical To Use AI?

In today's world, the ethics of using AI often come under scrutiny. However, to understand this better, let's consider some familiar analogies:

Is it ethical for your CPA to use spreadsheets and calculators instead of adding up your taxes by hand? Yes, of course. Spreadsheets and calculators enhance accuracy and efficiency, allowing CPAs to focus on more complex tasks and provide better service.

Is it ethical for you to drive across town instead of walking? Yes, of course. Cars are tools that save time and energy, enabling people to reach destinations faster and more efficiently. They are simply an evolution from walking to horseback riding to motor vehicles.

Is it ethical to use a vacuum cleaner instead of picking up dirt by hand? Yes, of course. Vacuum cleaners make cleaning more efficient and thorough, allowing for better hygiene with less effort.

AI is just the latest evolution in technology. It's designed to assist us, enhance our productivity, and enable us to achieve more. AI is here to stay, and we can use it to write for us, draw for us, speak for us, and more. The key is to use it the right way. It is just a tool, much like a nail gun.

Consider a nail gun. Is it ethical to use a nail gun? Yes, when used to build a house for the homeless, it's an incredibly positive use of technology. However, if it is used to harm someone, it's unethical. The tool itself is neutral; the morality lies in how it is used.

Using AI ethically means employing it to enhance our capabilities without misrepresentation or harm. Here are some practical guidelines and examples from my life:

- Writing: I use AI to draft content, but I ensure that the final version reflects my voice and ideas. It's a starting point, not the end product. For instance, when writing this book, I used AI to brainstorm ideas and organize my thoughts, but

every word published went through my personal review and edits to maintain authenticity.
- Design: AI helps me generate initial design concepts, which I then refine. This process is similar to using design software to create graphics instead of hand-drawing every element.
- Communication: AI can draft emails or social media posts based on my style, saving time while ensuring consistency. However, I always review these drafts to ensure they align with my intent and message.
- Customer Interaction: AI chatbots on my websites engage visitors in real-time, answering common queries and capturing lead information. This ensures no visitor is left unattended, much like a greeter in a retail store. However, complex issues are always escalated to human staff to provide personalized assistance.

The ethical use of AI is about transparency, accuracy, and maintaining a human touch where it matters most. It's about leveraging technology to enhance our capabilities without losing sight of our values and responsibilities.

In the following chapter, you will learn how to train AI to sound like you. This involves feeding it with your unique style, preferences, and anecdotes, ensuring the AI-generated content is a true extension of your voice. Remember, the goal is not to plagiarize or deceive but to enhance and streamline your efforts ethically.

By understanding and applying these principles, you can harness the power of AI effectively and responsibly, ensuring that technology serves as a tool for positive growth and innovation.

Train Your AI To Be YOU (Do This First!)

People are smart. We can tell when someone has used AI to write a LinkedIn post. Here are some dead giveaways:

- Title Case Text
- Overuse of emojis
- Overused words like "unlock," "unleash," "delve," "dive," and "supercharge"
- Overly formal or consistent tone
- Lack of personal touch and anecdotes

Because of this, you *must* take the time to train your AI tools to communicate in your unique voice. If you don't do this first, implementing the rest of the tactics in this book will backfire on you.

This chapter will guide you through the process of setting up and fine-tuning ChatGPT to sound just like you, enabling you to effectively clone yourself and amplify your impact.

Please do not go any further in this book until you have completed the exercises in this chapter. This foundation is needed for everything else to work correctly. If you skip this step, you will be doing yourself more harm than good.

WHY CLONE YOURSELF?

Before diving into the technicalities, it's important to understand the benefits of cloning yourself using AI. As a business leader or entrepreneur, your time is one of your most valuable assets. By training AI to replicate your communication style, you can:

- Scale your communication: Engage with clients, customers, and team members without being physically present.

- Maintain consistency: Ensure that your brand's voice remains consistent across all platforms and interactions.
- Increase productivity: Delegate routine tasks and focus on strategic decisions and creative work.

SETTING UP CUSTOM INSTRUCTIONS

The first step in cloning yourself is to set up custom instructions for ChatGPT. Custom instructions allow you to define how ChatGPT should respond in different scenarios, mimicking your tone, style, and preferences.

Step 1: Create a Google Doc for Custom Instructions

1. Start by creating a Google Doc where you can draft and save different sets of custom instructions. This is particularly useful if you manage multiple brands or need to switch between different communication styles.
2. This document should have two main sections for each brand voice: "About Me," and "My Writing Style"

Step 2: Write the "About Me" section

To make ChatGPT truly reflective of your voice, you need to provide it with detailed information about yourself in the custom instructions. When you see the prompt, "What would you like ChatGPT to know about you to provide better responses?" take the opportunity to share your interests, life goals, and any personal nuances that define you. Here are some examples of what you can include:

1. Choose the main character you want ChatGPT to sound like–you, your company, or a company mascot? Pick <u>one</u>.
2. If it's you...let's assume you own a business, and you want ChatGPT to speak from your personal voice as CEO. Your "About Me" section should contain the following:
    - My name is...
    - I am the CEO of...
    - My company does X

- Our three unique value propositions are...
- Our mission is...
- Our core values are...
- I really enjoy... (Mention hobbies, passions, and any areas of expertise.)
- Some of my personal goals are... (Share your long-term aspirations and what drives you.)

3. If you don't want to speak from your personal voice, and instead from the company's voice, then tweak the content like this:
    - Our company name is...
    - We do X
    - We've been in business since...
    - Our three unique value propositions are...
    - Our mission is...
    - Our core values are...
    - Our ideal customers are...
    - We solve this problem for our customers...
4. The more information you provide the better. Keep your total "About Me" area to 1500 characters.

While you certainly don't have to, I recommend that you speak from a human perspective. Most successful brands have a real or fictitious "attractive character" that is the spokesperson of the brand.

Here are some famous examples:

Character	Brand	Brand Voice Description
Flo	Progressive Insurance	Friendly, quirky, and helpful. Flo's cheerful and enthusiastic personality makes insurance approachable.
Gecko	Geico	Witty, charming, and straightforward. The Geico Gecko uses humor and simplicity to make insurance memorable.
Wendy	Wendy's	Bold, sassy, and humorous. Wendy's voice is known for its sharp wit and engaging social media presence.
Mickey Mouse	Disney	Wholesome, cheerful, and optimistic. Mickey embodies the magic and joy of Disney.
Steve Jobs	Apple	Visionary, innovative, and inspiring. Jobs' voice focused on simplicity, elegance, and groundbreaking technology.
Gary Vee	Vayner Media	Energetic, motivational, and candid. Gary Vee's voice is raw, honest, and driven by a passion for entrepreneurship.

Elon Musk	Tesla & Space X	Visionary, daring, and unconventional. Musk's voice is centered around innovation and pushing the boundaries of technology.
Jeff Bezos	Amazon	Customer-focused, ambitious, and strategic. Bezos' voice emphasizes efficiency, innovation, and customer satisfaction.
Colonel Sanders	KFC	Southern, charming, and folksy. Colonel Sanders' voice evokes a sense of tradition and home-cooked goodness.
Ronald McDonald	McDonald's	Fun, friendly, and approachable. Ronald McDonald's voice is all about family-friendly fun and happiness.
Tony the Tiger	Kellogg's Frosted Flakes	Energetic, positive, and encouraging. Tony's catchphrase "They're Grrreat!" is motivational and upbeat.
Mr. Clean	Mr. Clean	Strong, reliable, and tidy. Mr. Clean's voice conveys trustworthiness and the promise of a spotless clean.
Old Spice Guy	Old Spice	Confident, humorous, and over-the-top. The Old Spice Guy's voice is playful and exudes masculinity.

There are countless other examples, but you get the point. Your brand needs a character with a definitive voice.

In most small businesses, the marketing is written by the owner-operator anyway, so there is a chance that your brand voice and your personal voice are already very similar.

But if this is not the case, and if you already have an existing brand with it's own voice, start a new chat and use this prompt in ChatGPT:

> Read the website [your URL] and provide a description of the business
>
> *(wait for response)*
>
> Great. Rewrite your above description, but from the company's perspective, speaking as if you are [name of company] (use the term "we").

This will give you content to help fill out the "About Me" section of your document for an existing business with a brand voice.

Step 3: Create the "My Writing Style" section

Collect excerpts of your writing, such as blog posts, emails, or social media updates. Paste them all into your document. If you have a team member who writes for you, ask them to provide samples.

This will serve as the basis for ChatGPT to analyze and learn your style.

If you don't have a lot of written media, you can use transcripts of speaking you've done. Maybe you have appeared on podcasts or TV shows or YouTube videos. Google "free transcription service" to find a tool you like. At this time, my favorites are Riverside.fm/transcription (free) and Rev.com (paid).

Use ChatGPT to analyze the writing samples. You can do this with the following prompt:

> I'm going to feed you some excerpts from my writing and speaking, and I want you to analyze my writing and speaking style. Then I want you to craft a prompt for my custom instructions, that will help you speak in my voice. Do you understand?

Wait for a response, then feed it your examples by copying and pasting them into the prompt area. Alternatively, you can provide URLs to online content that represents your style. This feature requires the Plus version.

When you are done feeding it your excerpts, enter this prompt in the same chat:

> Great. Now write the custom instructions for ChatGPT to emulate my speaking style.

Yes, you are essentially telling ChatGPT to write instructions for ChatGPT.

Take those instructions and copy/paste them into the "My Writing Style" section of your document.

Now you can tailor the instructions further. Include details about your preferred tone, formality level, and any specific phrases or jargon you commonly use. Some things you may want to add to the description:

- Tone: Friendly? Professional? Witty? Humorous?
- Formality: Conversational? Formal?
- Grade Level: You can include commands like "Write on an upper high school or early college level," or "Speak like Gen Z." Just make sure it sounds authentically like you or your brand.
- Specific Phrases: Do you often say certain things, like "Let's circle back," "Absolutely," or "In a nutshell?" Include your commonly-used phrases in your instructions.
- Words To Never Use: It may be good to exclude certain words that are overused by AI, like "unlock," "unleash," "delve," "dive," and "supercharge"

- Celebrity Voice: You could ask ChatGPT "which famous figure is most similar to my writing/speaking style?" This is a quick hack to include to refine the voice. For me, it said Tony Robbins, Simon Sinek, or Gary Vaynerchuk. Then tell it to include a touch of that person's speaking style.

The more information you include in your description, the better. Every command will help your outputs sound more like you and less like AI.

Step 4: Enter Custom Instructions in ChatGPT

Access the custom instructions feature in ChatGPT by clicking on your profile name in the corner and selecting "Customize ChatGPT." Enter your instructions in the appropriate fields:

- Top Box: Information about your company and yourself.
- Bottom Box: Detailed writing instructions.

Tailor the outputs, If you find that ChatGPT's responses are too lengthy or wordy, you can refine the instructions further. Specify the desired reading level (e.g., 12th grade, 8th grade) to ensure the outputs are concise and appropriate for your audience.

Bottom line: keep tweaking the custom instructions until ChatGPT sounds like you.

Avoid the Rookie Mistake

Readers can tell when you have used AI out of the box. This is a rookie mistake. When AI-generated content lacks customization, it often feels generic and disconnected. Customizing your AI tools will make your content sound authentically yours.

Ensuring Authenticity

To ensure your AI-generated content is indistinguishable from human-written content, run your content through a plagiarism checker and an AI content checker. When I do this, my content usually comes back in the range of 80-98% human-generated, even though I used ChatGPT as my ghostwriter.

Here's how to do it:

1. **Use Plagiarism Checkers**: Tools like https://plagiarismdetector.net/ can help ensure your content is original and hasn't unintentionally mirrored existing texts. If some plagiarism is detected, services like Quillbot or Grammarly offer free paraphrasing tools.
2. **Employ AI Content Checkers**: Services like https://writer.com/ai-content-detector/ can analyze your text and provide a percentage score of human versus AI-generated content. If your content comes back as too AI-generated, rewrite it and go enhance your custom instructions with even more detail.
3. **Regular Reviews and Edits**: Continuously review and edit the AI-generated content to infuse more of your unique voice and ensure it aligns with your brand.

Practical Applications

Once you have set up custom instructions, you can leverage your cloned AI in various ways to boost your business:

1. **Content Creation**: Generate blog posts, social media updates, newsletters, and other content that aligns with your brand's voice.
2. **Customer Support**: Use AI to handle routine customer inquiries, providing quick and consistent responses.
3. **Internal Communications**: Streamline internal communications by having AI draft memos, emails, and reports that reflect your style.
4. **Marketing Campaigns**: Develop personalized marketing messages and campaigns that resonate with your target audience.

You are now equipped to do all of the above and so much more that we will teach in later chapters.

Tips for Success

- **Regular Updates**: Periodically update your custom instructions to reflect any changes in your style or brand voice.
- **Feedback Loop**: Review AI-generated outputs regularly and provide feedback to refine the instructions further.
- **Integration**: Integrate ChatGPT with other tools and platforms you use to maximize its utility and efficiency.
- **Occasionally Remind**: Sometimes you will need to remind ChatGPT to refer to your custom instructions and speak in your voice.

Cloning yourself using AI is a game-changer for business growth. By training ChatGPT to sound like you, you can extend your reach, maintain consistency, and boost productivity. Start by setting up custom instructions, gather writing samples, and continuously refine the outputs. With AI as your digital clone, you'll have more time to focus on what truly matters—driving your business forward.

Double Your Business: Just One More _____

Understanding the key drivers of your business is essential for growth. For every business, there are only three pivotal numbers that can significantly impact your revenue:

1. number of clients,
2. average order value (AOV), and
3. repeat purchase frequency

The concept of leveraging key business metrics to drive growth is not new. I learned it from the principles of Jay Abraham and Tony Robbins' Business Mastery. However, I took these foundational ideas further to create a simplified, actionable framework.

By strategically increasing each of these numbers by just 26%, you can double your revenue. This is what I call "The Rule of 26%."

Number of Clients

This metric refers to people who pay you for your product or service.

Quick side note: Notice that I call them "clients" instead of "customers."

"Customers" are people who just swing by, grab what they need, and move on. Being a customer is temporary and noncommittal.

But "clients"? They're the VIPs, the ones you build a relationship with, who trust you and come back for your expert advice and personalized service. Calling them "clients" not only sounds more elevated and sophisticated, but it also reflects a deeper, ongoing partnership. It's like the difference between a casual date and a long-term relationship – one's just a transaction, while the other's a journey. So, let's treat our business connections with the respect they deserve and call them clients.

This core metric has two parts:

1. **Acquire more leads.** Assuming you already have a working funnel, you're already converting a certain percentage of your leads (potential clients) into actual paying clients. So logically, the more leads you acquire, the more clients you will have.
2. **Increase conversion rate.** Your conversion rate is the percentage of leads that actually make a purchase. The higher this rate, the better.

Average Order Value (AOV)

AOV is the average amount a customer spends per transaction. By increasing this number, you maximize the revenue generated from each client. This can be achieved through upselling, cross-selling, and enhancing the perceived value of your offerings.

Repeat Purchase Frequency

Repeat purchase frequency refers to how often your clients return to buy from you. Encouraging repeat business is crucial for sustained revenue growth. Loyal customers tend to spend more and are more likely to refer others to your business.

The Power of Incremental Increases

To illustrate the impact of the Rule of 26%, let's consider an example. Suppose your business currently has:

- 100 clients
- An AOV of $10,000
- An average repeat purchase frequency of 2

This would result in a total revenue of $2,000,000:
100 clients × $10,000 AOV × 2 purchases per year = $2,000,000 revenue

Now, if you increase each of these metrics by 26%, you get:

- 126 clients
- An AOV of $12,600
- An average repeat purchase frequency of 2.52

This would result in a new total revenue of $4,000,000.

126 clients × $12,600 AOV × 2.52 purchases per year = $4,000,000 revenue

So increasing these numbers by 26% doesn't just grow your business by 26%... it doubles your revenue!

While the math is simple, looking at it this way helps us reframe business growth into something very manageable rather than daunting. Consider these scenarios:

- New Clients: If your business currently acquires 4 customers a day, ask yourself, "How hard is it to get just one more?"
- Average Order Value: If the average client spends $400, think, "How can I get them to spend just $100 more through upsells and adding value?"
- Repeat Purchase Frequency: If the average client shops with you 4 times, consider, "How can I encourage them to come back just one more time?"

This mindset shift makes the goal seem attainable and less overwhelming.

Strategies for Achieving the 26% Increase

Achieving a 26% increase in each of these areas is entirely feasible with the right strategies. Here are a few approaches to consider:

Increasing New Clients

Here are some ways to increase your number of leads, because the more leads you have, the more you can convert into clients:

- Ask your existing clients for referrals: Leverage your current satisfied customers to bring in new business.
- Over-deliver with impeccable service: Create "raving fans" who naturally promote your business.
- Social media marketing: Utilize platforms like Facebook, Instagram, LinkedIn, and Twitter to reach a wider audience.
- Referrals and word of mouth: Encourage happy clients to spread the word about your business.

- Content marketing: Create valuable content through blogs, YouTube videos, and newsjacking to attract potential clients.
- Start a podcast: Share your expertise and attract a dedicated following.
- Be a guest on podcasts: Reach new audiences by appearing on other people's podcasts.
- Trade shows: Showcase your products and services to a targeted audience.
- Case studies: Demonstrate your success through detailed case studies.
- Buy leads: Purchase leads from reputable sources to expand your potential client base.
- Cold calling: Directly reach out to potential clients.
- Networking events: Attend events to connect with potential clients and partners.
- Surveys: Gather insights and leads through targeted surveys.
- Billboards: Use high-visibility advertising to attract attention.
- Press releases: Announce significant business milestones to attract media coverage.
- Advertising: Invest in online and offline advertising to reach new customers.
- Partnerships: Form strategic partnerships to expand your reach.
- Giveaways: Attract attention and generate leads through contests and giveaways.
- Sponsorships: Sponsor events and initiatives to increase brand visibility.
- Public speaking: Establish yourself as an expert by speaking at industry events.
- Salespeople: Employ a dedicated sales team to pursue new leads.
- Online reviews: Encourage satisfied clients to leave positive reviews.
- Organic search and SEO: Optimize your website to attract organic traffic.
- Articles: Write and publish articles in industry publications.
- Radio: Advertise on radio stations to reach a broad audience.
- Volunteering and community events: Engage with your community to build relationships.
- Industry publications: Get featured in relevant magazines and journals.
- Direct mail: Send targeted direct mail campaigns.
- Webinars: Host informative webinars to attract and educate potential clients.
- Walk-ins: Create a welcoming environment for potential clients to visit.
- Influencer marketing: Partner with influencers to reach their followers.
- Testimonials: Showcase positive feedback from satisfied clients.

- Search and social media paid ads: Invest in targeted paid advertising.
- Collaborations: Collaborate with other businesses to expand your reach.
- Virtual and in-person events: Host events to connect with potential clients.
- Contests: Run contests to engage and attract new leads.

And there are so many more ways. Have you tried everything on this list?

If you seriously implemented one or two things from this list, do you think you could get just <u>one more</u> paying client for every four you already have? Simple, right? There's your 26%. (I know it is technically 25%, but remember we are reframing things to make them feel within reach. It's all about mindset first.)

Use the above list as a reference, and start executing to increase your number of new leads. Better yet, have AI tools help you execute. This is exactly what we will be learning in the next chapters.

Converting Leads into Paying Clients

There is a saying that "fortune is in the follow-up." Often, business owners make one sales attempt, then if there is no response, the prospect never hears from them again. Follow up with your leads regularly and often, at least seven to eleven times before you give up. After that, if you still don't get a response, move them to a different follow-up campaign. You can automate much of this using email marketing automation software like Mailchimp, ActiveCampaign, and Keap (formerly Infusionsoft). Make sure you are following the rules and being SPAM compliant.

Email is not the only way. Use several different touchpoints to stay in front of your prospects. Here is a non-comprehensive list:

- Email
- Phone calls
- Voicemail drops
- Text messaging
- Direct messaging (LinkedIn, Facebook, Instagram)
- Retargeting with paid ads
- Direct mail

- Promotional items
- Invite them to events (virtual and in-person)
- And more...

You'll want to follow up with any declines, cancels, abandoned carts, etc. and convert them to solid cash. However, no matter which of the above methods you use, make sure you are delivering valuable information to your prospect and not just selling all the time. According to Gary Vee, a good ratio for content to selling is 3:1—give, give, give, ask—or as Gary says, "Jab, jab, jab, right hook."

Have you tried everything on this list?

If you seriously implemented one or two things from the above list, do you think you could convert just <u>one more</u> lead into a paying client for every 4 leads you already have?

Use the above list as a reference, and start executing to convert more leads into clients. Or even better—you guessed it—we will teach you how to use AI and automation tools to help you execute some of these tasks faster.

Increasing Average Order Value

I learned from Verne Harnish, founder of Entrepreneurs' Organization (EO), that one often-overlooked strategy for increasing revenue, profitability, and cash, while also improving the perception of your brand, is to simply raise your prices. Here are some ways you can do it without causing a big fuss:

- Add more value (upsells and add-ons): Offer additional products or services that complement the original purchase.
- Clearly illustrate the return on investment: Show clients how spending more benefits them.
- Offer a guarantee: Provide a money-back guarantee to reduce the perceived risk.
- Give more, and receive more: Enhance your offerings to justify higher prices.
- Offer skip-the-line fees: Charge extra for priority service.
- Offer a VIP service level: Create premium packages with exclusive benefits.
- Warranties: Offer extended warranties for an additional fee.
- Subscriptions: Introduce subscription models for recurring revenue.

Have you tried implementing everything on this list? No? Then you have room to grow.

Do you think you can implement something from this list to get clients to spend just $1.04 more with you for every $4.00 they are already spending with you? There's your 26%.

Increasing Repeat Purchase Frequency

Here are some ways to increase how often your clients come back and shop with you:

- Suggest new and related products to existing clients: Introduce them to complementary products.
- Create a special offer or coupon for existing clients: Encourage repeat business with exclusive deals.
- Implement a membership or subscription service: Provide ongoing value through a subscription model.
- Regularly deliver valuable learning content via email and social media: Stay top-of-mind by offering continuous value.

And so much more. If a client shops with you an average of four times per year, do you think you can implement something from this list to get that client to shop with you one more time? There's your 26%.

Looking Ahead

In the following chapters, we will walk through specific strategies and practical steps to achieve these 26% increases using AI and automation. These tools can significantly enhance your efforts, making the process more efficient and scalable.

The Rule of 26% is a powerful framework for doubling your business revenue. By focusing on incremental improvements in new clients, average order value, and repeat purchase frequency, you can achieve substantial growth. Stay tuned as we explore the transformative power of AI and automation in driving these increases and propelling your business forward.

Prompt Engineering 101

Prompt engineering is the art and science of crafting precise and effective prompts to guide AI models like ChatGPT. This process maximizes the efficiency and relevance of AI-generated content, making it essential for various tasks from simple information retrieval to complex creative writing.

One of ChatGPT's most powerful features is its ability to assume any role or persona. Whether you need a chief marketing officer, a professional copywriter, or a notable historical figure, ChatGPT can emulate the tone, expertise, and perspective of that role. This versatility allows for a wide range of applications, from generating marketing strategies to drafting speeches or providing historical insights.

The "RACE" Method for Writing Prompts

The "RACE" method is a practical framework to ensure your prompts are clear, concise, and structured to get the best possible response from AI. This method stands for Role, Action, Context, and Execute.[4]

Role defines the persona or professional identity that ChatGPT should adopt. Specifying the role sets the stage for the AI to generate responses with the appropriate tone and expertise. Example: "Act as a financial advisor with expertise in retirement planning."

Action specifies the task or activity that you want ChatGPT to perform. This directs the AI to take a specific course of action, ensuring that the response is purposeful and aligned with your needs. Example: "Provide a detailed investment plan for someone looking to retire in 20 years."

Context provides background information and relevant details that the AI needs to consider while performing the action. This ensures that the response is tailored to the

---

[4] Michael Stelzner, "Prompt Engineering Fundamentals: How to Get Better Results With AI," *Social Media Examiner*, May 28, 2024 https://www.socialmediaexaminer.com/prompt-engineering-fundamentals-how-to-get-better-results-with-ai/

specific situation and includes all necessary nuances. Example: "For a client who is 45 years old, has a moderate risk tolerance, and currently has $200,000 in savings." This is also where you can include tone, writing style, etc. As long as they are enabled for new chats, your custom instructions that we created earlier provide additional context behind the scenes.

Execute describes the desired format or specific instructions on how to present the response. This could include the length, style, or any other formatting requirements. Example: "Write the plan in a concise report format, with sections for risk assessment, recommended investments, and a timeline for periodic reviews."

Leveraging ChatGPT's Role-Playing Capability

By leveraging ChatGPT's ability to adopt any role, you can enhance your business strategies, content creation, and problem-solving approaches.

Here's one prompt example that we will use later in the book:

> Act as my Chief Marketing Officer. Based on what you know about me, list relevant trade shows and industry events I could attend, and tell me the best ways to leverage my attendance/participation at each event. Include links for each event.

Let's break it down:

Role: "Act as my Chief Marketing Officer."

Action: "list relevant trade shows and industry events I could attend, and tell me the best ways to leverage my attendance/participation at each event."

Context: "Based on what you know about me" (this tells ChatGPT to refer to the memory and custom instructions)

Execution: "Include links for each event." I could have also included any other details about the way I wanted this formatted.

Mastering prompt engineering through the RACE method allows you to fully utilize ChatGPT's capabilities, ensuring that you get high-quality, relevant, and actionable responses. Whether you're drafting a marketing plan, seeking expert advice, or creating engaging content, the RACE method provides a structured approach to getting the most out of your AI interactions.

By applying these principles, you can streamline your workflow, enhance your creative output, and drive better results in your business endeavors. No programming knowledge needed.

Food for thought: What if you asked instructions of your *human* counterparts with this much clarity? Might you get different results?

Tools To Increase Number of Leads

This chapter is dedicated to equipping you with the tools and strategies necessary to attract a steady stream of potential customers. We'll delve into the latest AI-powered tools and automation techniques that can help you expand your reach, engage with a broader audience, and ultimately increase your lead generation efforts. From social media strategies to content marketing hacks, we've got you covered.

Imagine having a system in place that consistently brings in new leads while you focus on other aspects of your business. That's the power of leveraging modern technology to boost your lead generation. Drawing from real-world examples and personal anecdotes, I'll show you how to implement these tools effectively. We'll cover everything from harnessing the power of social media and SEO to utilizing AI for personalized marketing campaigns. By the end of this chapter, you'll have a robust toolkit to drive traffic and attract potential clients, setting the stage for significant business growth. Including detailed prompts and AI tools for each step, you'll be able to put these strategies into action immediately.

DAY 1: HARO HACK—GET FREE MEDIA COVERAGE AND CONTENT

HARO, or Help a Reporter Out, is basically like a matchmaking service for journalists and sources. It's where reporters go when they need expert quotes, and where savvy business owners and thought leaders (like you and me) go to get free publicity. Imagine a high-speed networking event where you can share your expertise, all while sipping your morning coffee. If you're looking to get your name and business in the spotlight with free PR, HARO is your golden ticket.

1. Go to https://www.connectively.us/ (formerly helpareporter.com also referred to as HARO)
2. Sign up as a Subject Matter Expert
3. Choose topics that relate to your expertise
4. Find a Query that you'd like to submit an article for
5. Enter this prompt in ChatGPT:

> Write a pitch for this HARO query for me, in my voice:
>
> [Copy/paste the entire query from HARO here]
>
> Mention [insert your personal experience, anecdote, or expertise here]
> Also mention [insert another personal experience, anecdote, or expertise here]
> Also mention [insert another personal experience, anecdote, or expertise here], and anything else you know about me that you think is a relevant fit for this query.

6. Then follow the instructions from the journalist to submit the pitch.
7. After submitting the pitch, whether you hear back from the journalist or not, go ahead and have ChatGPT write the full article by using this prompt:

> Great! Now act as an expert copywriter. Write the full article for this query, being sure to use my voice and writing style. Imagine that your article will appear in popular business publications.

8. Run the article through an AI checker, a plagiarism checker, and make sure everything is accurate and true. Read through it to make sure it actually sounds like you.

   AI content detector: https://writer.com/ai-content-detector/
   Plagiarism detector: https://plagiarismdetector.net/

9. Refine as necessary by either editing the prompt, asking ChatGPT to make changes, or just editing the output on our own.

Don't get your hopes up. You may never hear back from the reporter. But sometimes you will actually get published! Either way, it is a win for you, because now you have a written piece of content that you know people want to hear about, and you can repurpose this content for many other things, as we will outline in the next steps.

DAY 2 : PRESS RELEASES

Press releases are a fantastic way to get your business noticed by the media, boosting both your visibility and credibility. Thanks to AI, creating and distributing press releases has become easier than ever, enabling you to quickly produce high-quality, newsworthy content that grabs attention and drives engagement. Here's how:

1. In the same chat thread from above, type this prompt:

> Great! Now write a press release for me, using the article as inspiration. Here is a quick summary of something newsworthy happening in my business that is related to the article. Flesh it out into a full press release, formatted according to proper press release standards. Make up a quote from me to include in the release.
>
> Summary: [include your summary here]

2. Run the article through an AI checker, a plagiarism checker, and make sure everything is accurate and true. Read through it to make sure it actually sounds like you.
3. Refine as necessary by either editing the prompt, asking ChatGPT to make changes, or just editing the output on our own.
4. Go to Fiverr.com and search for "publish press releases"
5. Order a Fiverr gig with lots of lots of 5-star reviews—preferably one that submits to lots of reputable PR sites and/or to a well-known site like Yahoo! Finance—and follow the instructions to submit your press release. Prices could range anywhere from $5 to a few hundred dollars.

DAY 3 : BLOG POSTS

Blogging is not dead. It is still crucial for SEO and authority building. Here are the steps to automatically create blog posts about the topic we've started exploring:

1. Continuing in the same chat thread from above, now type this prompt:

> Now write a 500-word blog post on this topic for (your website). Start the title with "How..." Do not include subheadings. Always communicate what benefits are in it for the reader. Use an inspirational tone that shows the from-to journey for the reader. Use the pain-agitate-solution framework where appropriate. Make it friendly, in my voice. Keep it grounded and on a 12th grade level. Include a story about how we recently (briefly describe the story). End with a call to action for the reader to (buy product, call you, leave a comment, etc.).

2. Refine as necessary by either editing the prompt, asking ChatGPT to make changes, or just editing the output on our own.
3. Publish the post to your blog/website.

DAY 4: EMAIL NEWSLETTERS

Email newsletters are a powerful way to stay connected with your audience, drive engagement, and nurture leads into paying clients. With AI, creating and personalizing these newsletters has never been easier, allowing you to craft targeted, compelling content that resonates with your readers and boosts conversion rates effortlessly.

1. Type this prompt into the same chat thread from above:

> Now use the above to draft an email newsletter that we can broadcast to our subscribers. The subject line should be personalized, witty, sentence case, with no emojis and should evoke curiosity so people are intrigued to read what's inside.

2. Refine as necessary by either editing the prompt, asking ChatGPT to make changes, or just editing the output on our own.
3. Use the output in your email newsletter software and send the email.

DAY 5 : LINKEDIN ARTICLES

LinkedIn Articles offer a powerful platform to showcase your expertise, build credibility, and reach a professional audience. With AI tools, creating and publishing insightful, engaging articles is easier than ever, allowing you to focus on delivering valuable content without getting bogged down by the writing process.

1. Type this prompt into the same chat thread from above:

> Use all of the content from above to write a 1500-word Linkedin article. Also suggest a good image to use to support the article, then actually create that image.

2. Run the article through an AI checker, a plagiarism checker, and make sure everything is accurate and true. Read through it to make sure it actually sounds like you.
3. Refine as necessary by either editing the prompt, asking ChatGPT to make changes, or just editing the output on our own.
4. Post the article and image to LinkedIn

DAY 6 : MEDIUM ARTICLES

Medium.com is an online publishing platform where people share ideas, stories, and expertise with a wide, engaged audience.

Publishing on Medium.com offers several benefits for your business:

- Reach a Broad Audience: Medium has a massive user base, giving you the chance to attract new customers, partners, and investors.
- Boost Credibility: Publishing quality content on Medium can position you as an industry authority, building trust with your audience.
- Improve SEO: Medium ranks well on search engines, helping your articles appear in search results and driving organic traffic.
- Engage with the Community: Medium allows for interaction through comments and feedback, fostering a dynamic relationship with readers.
- Leverage Distribution Tools: Medium's tags and publications help your content reach the right audience, maximizing impact.
- Cross-Promote Content: Share your Medium articles on other social media platforms to increase traffic and engagement.
- Monetization Options: Medium's Partner Program allows you to earn money based on article engagement.
- Showcase Innovation: Share your forward-thinking ideas and strategies, inspiring and educating others while solidifying your position as a leader.
- Build Brand Awareness: Consistently publishing high-quality content increases brand recognition and trust.
- Network and Connect: Medium attracts professionals from various industries, providing opportunities to expand your network and collaborate.

Publishing on Medium can be a powerful way to grow your influence, reach new audiences, and drive your business forward.

Ready to share your insights with the world?

Here are the steps:

1. Type this prompt into the same chat thread from above:

> Now write an article for Medium.com. Is there anything you would change?

2. Review for accuracy and refine as necessary by either editing the prompt, asking ChatGPT to make changes, or just editing the output on our own.
3. Publish the article on Medium. You will need to first set up an account if you don't already have one.

DAY 7 : GET FEATURED ON INFLUENTIAL NEWS SITES

Brand Push helps small businesses gain visibility and credibility. By leveraging their extensive network, Brand Push can get your business featured on influential news sites like Yahoo Finance, MSN, and Business Insider. This not only enhances your brand's online presence but also builds trust with potential customers.

For our businesses, utilizing Brand Push means more eyes on our innovative solutions and greater engagement from our target audience. It's like having a megaphone that amplifies our success stories and business strategies to a global audience, ensuring that we stand out in a crowded market.

Here are the steps:

1. Type this prompt into the same chat thread from above:

> Now act as an expert copywriter who is frequently published in major business publications.
>
> I would like to use https://www.brandpush.co/ to get featured on influential news sites. Using all of the content from above, craft an article that will perform well on major news outlets like Yahoo Finance, MSN, Business Insider, and more.

2. Review for accuracy and refine as necessary by either editing the prompt, asking ChatGPT to make changes, or just editing the output on our own.
3. Go to https://www.brandpush.co/ and choose the package you want, then order it. Follow the instructions on their site.

DAY 8 : ANSWER QUESTIONS ON QUORA

Using Quora in your content marketing plan can be highly effective for building brand awareness and driving traffic to your business. Here's a step-by-step guide:

Step 1: Create a Quora Account

- Sign Up: Register for a Quora account if you don't already have one.
- Optimize Your Profile: Use a professional photo, write a compelling bio, and include links to your business website and social media profiles. Highlight your expertise.

Step 2: Identify Relevant Topics

- Search Topics: Use Quora's search feature to find topics related to your expertise.
- Follow Topics: Follow these topics to stay updated on the latest questions and discussions.

Step 3: Provide Valuable Answers

- Find Questions: Look for questions that are relevant to your expertise and that your target audience might be asking.
- Write Thoughtful Answers: Provide detailed, valuable answers that address the question thoroughly. Use your failures, success stories, and insights from your real-world experiences. Now that your ChatGPT knows all about you and your writing style, use it to help you draft answers to the questions.
- Include Links: Where appropriate, include links to your blog posts, articles on Medium, or relevant pages on your website. Ensure these links provide additional value to the reader and aren't overly promotional.

Step 4: Engage with the Community

- Upvote and Comment: Upvote other helpful answers and leave comments to engage with the community.
- Follow Influencers: Follow influencers and thought leaders in your industry. Engaging with their content can help increase your visibility.

Step 5: Share Your Content

- Post Questions: Ask questions that prompt discussions relevant to your industry. This can help you understand your audience's needs and position you as a thought leader.
- Share Content: Occasionally share your blog posts or articles as answers to relevant questions. Ensure these posts offer valuable insights and aren't just promotional.

Step 6: Analyze and Adjust

- Track Performance: Use Quora's analytics to track the performance of your answers. Look at views, upvotes, and engagement to see what's working.
- Adjust Strategy: Based on the analytics, adjust your strategy to focus more on the types of questions and answers that are driving the most engagement and traffic.

Step 7: Leverage Quora Ads (Optional)

- Run Ads: Consider running Quora ads to target specific audiences interested in topics related to your business. This can help drive more traffic to your website and increase brand awareness.

Quora is a powerful tool for content marketing that can help you build brand authority, drive traffic, and engage with potential customers. By providing valuable answers and engaging with the community, you can position yourself as a leader in your field.

DAY 9 : X/TWEETS

Posting on X (formerly Twitter) is more than just a social media strategy; it's a dynamic way to engage with your audience in real-time. Businesses should leverage X to share insights, showcase innovations, and build authentic connections with their followers.

It's a platform where you can spark conversations, get immediate feedback, and stay on the pulse of industry trends. By consistently posting valuable content, you create a space for your brand to be seen as a thought leader.

Plus, the viral nature of X can amplify your message far beyond your immediate reach, opening doors to new opportunities and growth. If you want to drive engagement and build a strong, interactive community, X is where you need to be.

1. Type this prompt into the same chat thread from above:

> Extract good tweets from the articles, blog, and newsletter above. Give me 10 highly-quotable tweets that are thought-provoking, poetic, witty, and/or polarizing, to increase the likelihood of them going viral. Include a call to action at the end of each tweet. Write on a high school level. Do not use emojis or hashtags.

2. Review for accuracy and refine as necessary by either editing the prompt, asking ChatGPT to make changes, or just editing the output on our own.
3. Schedule the tweets natively on X.com (scheduling only works on desktop for now). Or using your favorite social media scheduling app. We will discuss these apps in detail later.

DAY 10: INSTAGRAM THREADS

Posting to Instagram Threads, even if you're already active on X/Twitter, is essential for maximizing your reach and engagement. Each platform has its unique audience and vibe. While X/Twitter is great for quick, real-time updates and broad discussions, Instagram Threads offers a more intimate, community-focused environment where your brand can foster deeper connections. By leveraging both, you ensure your message resonates across different user bases, tapping into varied engagement styles and preferences. It's about meeting your audience where they are, maximizing visibility, and driving growth through diversified digital touchpoints.

1. Type this prompt into the same chat thread from above:

> Would you change these for Instagram Threads? Or would you use the same tweets?

2. Review for accuracy and refine as necessary by either editing the prompt, asking ChatGPT to make changes, or just editing the output on our own.
3. Post natively on the Threads app, or threads.net, or schedule the posts using your favorite social media scheduling app. We will discuss these apps in detail later.

DAY 11: BATCH CREATE TEXT-IMAGE POSTS IN SECONDS

You've probably seen Instagram posts that are images of tweets, or inspirational quotes with a cool background photo. These are commonly referred to as "quote posts" or "text overlay posts," and can add some variety to your social media profile to mix up with videos and photos.

You can generate dozens of these in seconds with ChatGPT and Canva!

Canva's Bulk Create feature allows users to create designs in bulk, including images, on desktop. It's available on Canva Pro and other paid versions. Here are the steps to bulk create images in Canva:

1. Choose a template or design
2. Select Apps from the editor side panel
3. Search "Bulk create" and click on it
4. In ChatGPT, use this prompt, in the same chat from above, to generate your tweets:

> Act as a person who consistently goes viral on Twitter. Read my article that we wrote above, and extract as many viral tweets as you can and export them in a csv file.

5. Or, if you have written a book or article, you could say this:

> Act as a person who consistently goes viral on Twitter. Read this excerpt from my book and write as many viral tweets as you can and export them in a csv file: [paste your book text here]

6. Back in Canva, click Import data for Bulk Create
7. Connect data to your elements
8. Click Generate designs

You can do the same thing for quotes. Either quotes from famous people, or you can ask ChatGPT to help you come up with your own memorable quotes based on your content.

Now you have a bunch of images to include in your social media library for scheduling.

DAY 12: LIVE VIDEOS

Now that we've produced lots of written content on the subject, we will have ChatGPT direct some videos for us. It is best to start with live videos because they create real-time engagement with your audience, build authenticity, and allow for immediate feedback and interaction, making your content more dynamic and relatable.

I know some people are very hesitant to show their face on camera. I get it. I was too. Here are some tips for first-timers to help you get over that apprehension and go live:

- Practice Makes Perfect: Start by recording yourself in private to get comfortable with speaking on camera.
- Prepare Your Space: Choose a quiet, well-lit area free from distractions to help you feel more confident.
- Engage with Your Audience: Focus on having a conversation rather than delivering a monologue. This makes the experience more interactive and less intimidating.
- Be Yourself: Authenticity resonates with viewers. Don't worry about being perfect; just be genuine.
- Start Small: Begin with short live sessions to gradually build your confidence and comfort level.
- Use Prompts: Keep a few key points or prompts nearby to help you stay on track without feeling scripted.
- Breathe and Smile: Relaxation techniques like deep breathing can help calm nerves, and a smile can make a big difference in how you come across on camera.
- Focus on Serving Others: Remember, you're providing value to your audience. Shifting your focus to serving others can reduce nervousness because you're thinking less about yourself and more about how you can help.
- No One Cares That Much: Most people are concerned with themselves and their own lives. They're not scrutinizing you as closely as you might think, so don't sweat the small stuff.

Remember, the more you practice, the easier it will get. You've got this!

1. Type this prompt into the same chat thread from above:

> I'd like to go live on Facebook, Instagram, and YouTube about this subject matter. Please write out a bulleted script so I will have talking points for the live video. Optimize the script to be both educational and entertaining, to keep viewers engaged and watching. Include a catchy title for the live video, and give suggestions on the best background/environment.

2. Review for accuracy and refine as necessary by either editing the prompt, asking ChatGPT to make changes, or just editing the output on our own.
3. Use a tool like OneStream to go live on multiple platforms simultaneously. Or you could get three separate smartphones and a multiple phone tripod, then log in to each app on go live on the separate phones, at the same time.
4. Make sure you save the recording of your live video to use for later.

DAY 13: LONG-FORM VIDEOS

Long-form videos can be a catalyst for building trust and authority with your audience, offering deep insights and comprehensive content that shorter clips can't match. Thanks to AI, scripting and publishing these videos has never been easier, allowing you to focus on delivering value while AI handles the heavy lifting of content creation and optimization.

Follow these steps:

1. Publish the recording of your live video to YouTube, and/or
2. Ask ChatGPT to write a script for a long form YouTube video. Here is the prompt:

> I'd now like to record some long-form video content about this subject for my YouTube Channel. Act as a YouTube expert. Please write a title, script, and thumbnail idea that will result in more clicks, views, and retention.

3. Load the script into a teleprompter app. There are several available in the app stores if you are recording on a mobile device, and free ones online if you are recording from a desktop or laptop.
4. Record the video.
5. Edit the video (optional). My favorite easy editing apps are InShot, CapCut, and iMovie. Descript is another cool editing tool that allows you to edit a video by editing its text transcript.
6. Publish the video to your YouTube channel.

Bonus tip: Use Pictory.AI to automatically turn your script into a video, narrated with an AI voice and b-roll footage selected by AI to go along with the words.

DAY 14: AUTOMATED WEBINARS

I know what you're thinking: "Do I really need to add one more thing to my plate?" Trust me, I get it. But if there's one strategy that can genuinely move the needle for your business, it's hosting a webinar. Nail it once, and then automate it.

Let's break down why this is a no-brainer move:

- Build Trust and Authority
    - When you host a webinar, you're positioning yourself as an expert in your field. People buy from those they trust, and nothing builds trust faster than sharing your knowledge and expertise live. It's like giving your audience a backstage pass to your brain.
- Engage with Your Audience in Real-Time
    - Webinars allow for real-time interaction. You can answer questions, respond to comments, and engage with your audience in a way that's just not possible through email or social media. This interaction builds a deeper connection and keeps your audience engaged.
- Showcase Your Expertise and Value
    - A webinar is the perfect platform to showcase the value you bring to the table. Whether you're teaching a new concept, sharing industry insights, or demonstrating your product, you have an uninterrupted block of time to show your audience why you're the go-to person in your niche.
- Generate High-Quality Leads
    - People who sign up for webinars are often more engaged and interested in your content than those who just follow you on social media. They're willing to give up their time to listen to you, which means they're likely to be high-quality leads.
- Create Evergreen Content
    - Your webinar doesn't have to be a one-and-done deal. Record it and you've got evergreen content that you can share on your website, social media, or through your email list. This means your effort continues to pay off long after the webinar is over.

- Increase Sales
    - A well-executed webinar can drive serious sales. By the end of your presentation, you've built trust, provided value, and addressed objections. This primes your audience to take action, whether that's purchasing your product, signing up for your service, or whatever your call to action may be.
- Expand Your Reach
    - Webinars are a fantastic way to reach a global audience. You're no longer limited by geography. Anyone with an internet connection can join in, expanding your reach and potential customer base.
- Showcase Case Studies and Testimonials
    - Sharing real-life success stories during your webinar adds credibility and shows your audience that what you're offering actually works. This can be incredibly persuasive for those on the fence.
- Offer Real Value and Education
    - This isn't just about selling. Webinars are a great way to provide real value and education to your audience. When you focus on genuinely helping your audience, the sales will naturally follow.
- Boost Your Brand's Visibility
    - Hosting regular webinars can significantly boost your brand's visibility. As more people hear about your webinars and the value you provide, your brand recognition will grow, leading to more opportunities down the line.

So, if you haven't already, it's time to seriously consider adding webinars to your marketing strategy. It's one of the most effective ways to connect with your audience, build your brand, and drive sales.

Think webinars are outdated. Guess again. Webinars have been strong for decades, and still continue to be. What we call them just changes from time to time. Here are some trending alternatives to "webinar" right now:

- Virtual Workshop
- Live Masterclass
- Interactive Session
- Online Seminar

- Digital Summit
- Virtual Learning Event
- Live Training
- Online Bootcamp
- Virtual Experience
- Live Coaching Session
- Interactive Webinar
- Digital Conference
- Online Intensive
- Virtual Roundtable
- Live Stream Session

After doing a few live videos and long-form videos, you should be warmed up to do a webinar. Try this prompt to help you:

> Write a webinar script using Russell Brunson's Perfect Webinar format. What do you need from me to write the script?

Several software platforms offer automated webinar features, allowing you to host pre-recorded webinars with AI-driven scheduling, registration management, follow-ups, and attendee interactions. These platforms can help you convert attendees into leads effectively. Here are some popular options:

- EverWebinar:
  - Features: Automated webinars, scheduling, registration pages, email notifications, live chat simulations, analytics.
  - Benefits: Easy to set up, integrates with other marketing tools, robust automation features.
- WebinarJam:
  - Features: Live and automated webinars, flexible scheduling, registration management, automated emails, polls, and surveys.
  - Benefits: High engagement tools, comprehensive analytics, strong integration options.

- Demio:
  - Features: Automated webinars, registration pages, automated email sequences, interactive features like polls and Q&A.
  - Benefits: User-friendly interface, customizable branding, real-time engagement options.
- GoToWebinar:
  - Features: Pre-recorded and automated webinars, scheduling, registration management, follow-up emails, audience interaction tools.
  - Benefits: Reliable and well-known platform, strong analytics, integration with various CRM tools.
- StealthSeminar:
  - Features: Automated webinars, just-in-time scheduling, registration pages, email reminders, interaction tools.
  - Benefits: Advanced scheduling options, strong customization, detailed analytics.
- ClickMeeting:
  - Features: Automated and on-demand webinars, registration management, automated emails, interactive features.
  - Benefits: Easy to use, good for large audiences, strong analytics and reporting.
- EasyWebinar:
  - Features: Automated webinars, live chat, registration pages, automated email notifications, detailed analytics.
  - Benefits: Combines live and automated webinars, strong integration options, user-friendly.

DAY 15: PODCASTING THE EASY WAY

Launching a podcast can be daunting. I started and stopped for years.

But I believe every business or brand should have a podcast because it's a powerful way to build a deep connection with your audience and establish authority in your niche. Think about it – podcasts offer a unique opportunity to share your insights, stories, and expertise in a personal, engaging format that your audience can consume on the go. It's like having a one-on-one conversation with your listeners, building trust and loyalty over time. You are literally in your listener's ear while they are exercising, driving, cleaning up, or naked in the shower.

Plus, podcasts are an incredible tool for content marketing, allowing you to repurpose episodes into blog posts, social media content, and more. If you're looking to elevate your brand and reach a wider audience, a podcast is a no-brainer.

Now that you have long form video from the steps above, it can be super easy to produce audio content. Here are the steps:

1. Drop one of your long-form video files into Auphonic. It's free, and allows you to separate the audio from your video while also using AI to optimize the sound quality. Try it at https://auphonic.com/
2. If you don't already have the transcript, try Riverside.fm/transcription (free) or Rev.com (paid).
3. Use the transcript and ChatGPT to generate a title and show notes for the episode.

    I will share with you the exact prompt that I use to generate my podcast show notes, so you can tweak it with your podcast's details. I actually set up a custom GPT for this (we will go over that later).

Here is my prompt:

> Your role is to assist in creating blog posts based on Kingspiration podcast episodes. You'll receive transcripts and transform them into inspirational and educational show descriptions, written in a third-person, active voice. The posts should be concise, around 150 words, designed to evoke curiosity and make readers want to listen to the full episode. Don't summarize the whole episode; leave some things to curiosity. First come up with 5 curiosity-evoking titles to choose from, all titles should contain the word "you".
>
> Then your show description should begin with a thought-provoking question and a paragraph about the show, then include a section titled "What you will learn in this episode," with bullet points each starting with "How to...", highlighting tangible, actionable benefits.
>
> Also include a "Questions answered" section listing the questions that the episode answers (list the questions in bullet point, but do not give the answers). Conclude each post with a call-to-action (CTA) inviting readers to listen to the episode, subscribe, and leave a review. Initially, ask for the episode transcript to begin the task. Maintain an active, third-person narrative style throughout the post. Write on a high school level.
>
> Lastly, what are the most interesting sound bites from this episode? I need one sound bite to extract as the hook, the very first thing listeners hear, to draw them in and compel them to listen to the entire episode. Give time stamps.
>
> About the Kingspiration podcast: American entrepreneur Ethan King and South African entrepreneur Justin King share global insights and actionable tools for business growth, personal development, relationships, physical and mental health optimization. New episodes weekly. No fluff, just real experiences and raw advice. Subscribe now and never miss an episode.

4. Next, you'll need to choose a podcast hosting service if you don't already have one. I currently use Podbean. I hear really good things about Buzzsprout too. You can search Google or ask ChatGPT which podcast host is the best. Don't get hung up on this step. They all essentially do the same thing. Just pick one and keep going.
5. Once you have set up your account and your show on your podcast host, click New Episode.
6. Upload your audio file, insert your title and show notes, then publish or schedule the episode. Voila!

DAY 16: CREATE SHORT-FORM VIDEOS WITH LESS EFFORT

According to research from NP Digital, short-form videos generate more engagement than any other type of content.

Type Of Content That Generates The Most Engagement

Short Form Videos	Long Form Videos	Interviews, Podcasts, Expert Talks	Live Video	Memes	User Generated Content	Other
31.38%	15.51%	13.84%	12.37%	10.31%	10.12%	6.47%

Source: NP Digital - We analyzed 6,302,491 posts on social media.

You may be thinking, "well that's great, but I don't have time to record a bunch of videos" or "I don't even know what to say" or "I hate being on camera."

Here are some hacks to make it easy:

a. Talking Head Videos: Ask ChatGPT to write scripts for you for talking-head videos. Use the prompts below in the same thread where you already had it generate your written content. That way, it will weave your tips, anecdotes and experiences into the scripts, but in a way that will increase the likelihood of virality. The prompt:

> Now act as an expert on viral videos. You have analyzed millions of short-form videos and figured out the common elements that make them go viral. What did you discover?
>
> (wait for response)
>
> Great, now refer to my subject from earlier in this chat. Write 10 different talking-head video scripts related to the topic. Tell me exactly what to say in each video. Each video should incorporate all the elements of virality that you listed.

b. Use Interview Questions: Usually if someone asks you a question about your business origin, industry or expertise, you can probably talk about it for hours. The trick is, they have to be the right questions to get the best type of answers out of you. Try this prompt:

> Now I'd like to record some long-form video content about this subject. Please write questions for someone to interview me about this subject, as if I am a podcast guest. Include insightful questions that will evoke a TikTok-worthy answer from me.

c. Have someone else ask you the questions off-camera, while they record you answering the questions. You don't need a fancy camera because most modern smartphones come with a good camera built in. If you don't have someone else, just buy a kit from Amazon that has a tripod, mic, and light. Good sound and lighting are key. You'll look best on camera when light is shining toward your face (not from above your head, or behind you). For the exact kit that I recommend, go to https://EthanKing.com/kit

After you have the scripts, use a teleprompter app to make recording easier. Some I like are PromptSmart Pro and BigVu, but there are lots to choose from—try them out to see which ones you like best.

If recording on your computer, just do a web search for "free teleprompter;" there are a ton of them out there.

Either way, the words you are reading should be placed near the camera lens so that your eyes are looking toward the camera. This makes it feel like you are making eye contact with the viewer. Some video editing apps even have an eye contact feature, where it uses AI to make it seem like you are looking at the camera.

Don't overthink it! If you can read (which I assume you can because you are reading this book), you can do this!

DAY 17: PUBLISH SHORT-FORM VIDEOS ON AUTOPILOT

Now that you've laid the groundwork, here's a fun and easy step!

OpusClip turns long videos into shorts, and publishes them to all social platforms in one click. Here's how to use it:

1. Go to https://www.EthanKing.com/clips
2. Copy a link to your long-form YouTube video, and paste it into Opus where it says "Drop a video link." Then click "Get free clips." You'll probably need to set up a free account.
3. Let the AI work, and in a few minutes you will have short video clips for TikTok, Instagram Reels, Facebook, LinkedIn, and YouTube Shorts. The coolest part is that Opus uses AI to rank your videos for likelihood of virality!
4. From here, you can autopost to your different accounts with captions and everything written for you. Or you can download the clips to schedule them in your favorite app.

DAY 18: POST TO SOCIAL MEDIA STORIES DAILY

In the realm of social media, your "stories" are a dynamic way to engage your audience that differs significantly from regular posts. Let's break down what stories are, why they matter, what content works best, and how you can streamline your efforts with scheduling apps.

What Are Stories?

Stories are short, ephemeral pieces of content that disappear after 24 hours. Unlike regular posts, stories are designed to be quick, informal, and more engaging. They can include photos, videos, text, and interactive elements like polls and questions. Platforms like Instagram, Facebook, TikTok and Snapchat offer story features.

Why Post to Stories?

Posting to stories daily can significantly enhance your social media presence. If you don't have activity in your stories, some people won't even know you're there. It's like having your house porch light on. Here's why:

- **Higher Engagement:** Many users spend more time flipping through stories than scrolling through posts, as stories appear at the top of social media apps, catching the user's attention first.
- **Real-Time Interaction:** Stories offer real-time updates and a sense of immediacy that traditional posts can't match, making them perfect for sharing behind-the-scenes looks, spontaneous thoughts, or live events.
- **Enhanced Visibility:** Stories are often prioritized by social media algorithms, giving you more visibility and reach compared to regular posts.
- **Interactive Features:** Stories come with interactive tools like polls, quizzes, and Q&A stickers, allowing you to engage directly with your audience and gather valuable feedback.
- **Personal Connection:** They humanize your brand by offering a glimpse into the day-to-day operations, making your audience feel more connected and involved.

Good Types of Content for Stories

Creating engaging content for stories involves variety and creativity. Here are some effective types of content to share in your stories:

- **Behind-the-Scenes Content:** Show what goes on behind the scenes in your business to build trust and humanize your brand.
- **Interactive Polls and Quizzes:** Use the interactive features to create polls, quizzes, and Q&A sessions to engage your audience.
- **User-Generated Content:** Share content created by your customers or fans, such as photos of them using your product, to provide social proof and build community.
- **Product Teasers and Launches:** Share sneak peeks of upcoming products or exclusive launches to create anticipation and excitement.
- **Day-in-the-Life:** Document a day in the life of a team member or your business owner to offer a personal touch and deeper insight into your company culture.
- **Tutorials and How-Tos:** Create short, instructional content that shows how to use your products to add value to your audience and position you as an expert.
- **Flash Sales and Special Offers:** Announce limited-time offers, discounts, or flash sales to drive immediate action and boost sales.
- **Customer Testimonials and Reviews:** Share positive feedback from your customers to build credibility and trust by showing real customer experiences.
- **Event Coverage:** Share live updates from events, whether they are in-person or virtual, to keep your audience informed and make them feel part of the action.
- **Motivational Quotes and Messages:** Post inspirational quotes or messages relevant to your brand or industry to engage your audience and increase your reach.

Someone can reply to your stories and it will appear in your messages. Keep the conversation going! I've made plenty of sales through stories and subsequent direct messages.

Apps to Schedule Stories

Scheduling your stories can save you time and ensure consistent engagement. Here are some top apps that let you schedule stories across various platforms:

- **Later:** Allows you to visually plan and schedule Instagram stories, making it easy to manage your content calendar.
- **Buffer:** Offers a story planner and scheduling for Instagram Stories, along with analytics to track performance.
- **Hootsuite:** Provides comprehensive social media management, including story scheduling for Instagram, and offers robust analytics.
- **Planoly:** Designed specifically for Instagram, this app allows for easy drag-and-drop story planning and scheduling.
- **Tailwind:** Supports scheduling for Instagram Stories and Pinterest, and includes content creation tools and analytics.
- **Sprout Social:** Offers story scheduling for Instagram along with in-depth analytics and reporting to help track story performance.

By leveraging these tools, you can ensure that your stories are posted consistently, allowing you to maintain a steady stream of engagement with your audience without the need for manual posting.

DAY 19: VIRAL TIKTOK VIDEOS

TikTok's algorithm is a powerful tool designed to personalize your feed based on your interests. It sees what you stop to watch, and shows you more videos that are similar. TikTok analyzes user interactions, video information, and device/account settings to curate content that's most likely to engage you.

This means that the more you interact with certain types of content, the more tailored and relevant your feed becomes, making it easier for your videos to reach the right audience if you know how to tap into trending topics and engaging formats.

Because of this, TikTok can provide unprecedented reach and engagement. There are many stories of new users posting a video that goes viral and gets 1M views. With AI, scripting and publishing these videos has become easier than ever, allowing you to create compelling content that captures attention and drives traffic effortlessly.

In the same thread from earlier, try this prompt:

> Write 3 different viral TikTok video scripts for my company about the topic. Start with a strong hook that will make people stop scrolling, then teach something (start with "Here's how to" then give steps 1 2 3), then end with a call to action.

DAY 20: BE A FIRST-MOVER ON NEW SOCIAL MEDIA APPS

Did you know there are new social media apps that haven't been discovered by the masses but have hundreds of thousands of users? Posting on these platforms offers a unique advantage: you can be a big fish in a small pond.

Being an early adopter on a new social media platform can significantly boost your organic growth. Established platforms like Facebook, Instagram, and Twitter often require paid promotions to gain visibility. In contrast, new platforms are less saturated, providing a greater opportunity to grow organically and reach a wider audience without spending a fortune on ads.

When we first got on Facebook with stuff4GREEKS in the early days (circa 2008), we quickly grew to 100,000 fans, and they all saw our posts. But as Facebook grew, it began throttling visibility, forcing us to boost posts and run ads to maintain engagement. Getting in early on a platform can give you an edge before these changes occur.

Examples of Emerging Platforms

- Lemon8: With over 1.5 million users, Lemon8 focuses on visual content, combining elements of Instagram and Pinterest. It's a great platform for brands that rely heavily on aesthetic appeal .
- Clapper: A rapidly growing platform similar to TikTok but without the same level of competition, Clapper boasts over 500,000 users. It's an excellent place to reach a new audience through short videos .
- BeReal: This app encourages authenticity with its unique posting method. Users share unfiltered photos once a day when prompted. With millions of downloads, BeReal fosters a genuine connection with its audience .
- Vero: Vero offers an ad-free experience and a chronological feed, appealing to users frustrated with algorithmic timelines. It has seen a resurgence in popularity, especially among creatives and photographers.

And there are so many more. New platforms are always popping up. Just do a web search for "newest social media platforms (month and year)" to find the latest ones. These new apps provide fresh opportunities to reach untapped audiences.

These networks do come and go, but you never know which one will be the next big hit. By staying ahead of the curve and adopting new social media platforms early, you can position yourself for significant organic growth and build a strong presence before the masses arrive.

Since some of these are so new, there may not be a way to auto schedule posts to them yet. But since you have already done the heavy lifting in the steps above, just a quick copy/paste to share the same post could be worth it.

Prompt:

> What are some widely-used social media platforms that most people don't know about yet?

DAY 21: RESPOND TO COMMENTS AND REVIEWS

In the modern digital landscape, customer feedback is more accessible than ever. Comments and reviews flood in from social media platforms, websites, and review sites. How you respond to these reviews can make or break your business. This chapter will explore why it's crucial to engage with comments and reviews and how ChatGPT can help streamline and enhance this process. We will also provide some actionable prompts to get you started.

Customer feedback is a goldmine of information and an opportunity to build your brand reputation. Here's why responding to comments and reviews is vital:

- Builds Trust and Credibility: When you respond to reviews, you show customers that you value their opinions and are committed to improving their experience. This transparency fosters trust and strengthens your credibility.
    - Example: Consider a small café that diligently responds to every review, whether positive or negative. A customer leaves a glowing review about their excellent latte and service. The café owner responds with a personalized thank you, mentioning the customer's specific comment about the latte. This personal touch makes the customer feel valued and more likely to return.
- Improves Customer Loyalty: Addressing concerns and appreciating positive feedback can turn a one-time buyer into a loyal customer. Engaged customers feel a connection to your brand and are more likely to become repeat clients.
    - Example: A software company receives a complaint about a bug in their latest update. By promptly acknowledging the issue and providing a timeline for a fix, the company not only resolves the issue but also demonstrates their commitment to customer satisfaction. This can turn a frustrated customer into a brand advocate.
- Enhances SEO: Engaging with reviews can improve your search engine ranking. Regular updates and interactions signal to search engines that your business is active and customer-focused.

- Example: An e-commerce store consistently responds to customer reviews, incorporating relevant keywords naturally. Over time, this activity can boost their SEO efforts, leading to higher visibility on search engine results pages (SERPs).

To get you started, here are some prompts you can use with ChatGPT to generate effective responses to comments and reviews:

> Write a thank you response for this 5-star review and include relevant keywords to help boost our SEO: [paste the review here]

> [upload a screenshot of an Instagram post]
> I posted this on Instagram. Reply to this Instagram comment in a positive, but witty way [paste the comment here].

DAY 22: DEPLOY AI CHATBOTS FOR WEBSITE ENGAGEMENT

Imagine walking into a retail store and no one greets you. You'd probably feel unwelcome and might even leave without buying anything. Now, translate that experience to your website. Your AI chatbot is like the friendly greeter who ensures every visitor feels acknowledged, attended to, and guided.

AI chatbots are revolutionizing customer engagement by providing real-time interaction with website visitors. They can instantly answer queries, guide users to the information they need, and even help with transactions. This immediate assistance not only improves the user experience but also significantly boosts your conversion rates.

Key Benefits:

- 24/7 Availability: Your chatbot never needs a break, ensuring round-the-clock customer support.
- Instant Response: Reduces wait time for customers, leading to higher satisfaction.
- Scalability: Handles multiple conversations simultaneously, something a human agent cannot do.

Capture Lead Information for Follow-Up

Beyond just answering questions, chatbots are excellent tools for capturing lead information. By engaging visitors in conversation, they can gather essential details like names, email addresses, and preferences. This data is invaluable for your follow-up marketing campaigns, allowing you to nurture leads and convert them into paying customers.

Example: A visitor shows interest in a product but leaves without purchasing. Your chatbot can capture their contact details and preferences, enabling your sales team to follow up with a personalized offer, increasing the chances of closing the sale.

Here are some of the top-rated chatbots right now:

- **Intercom:** Known for its robust features and integrations, Intercom provides a seamless chat experience, helping businesses improve customer support and engagement.
- **Drift:** Focused on sales and marketing, Drift's chatbots are designed to qualify leads and book meetings automatically, streamlining the sales process.
- **Tidio:** Tidio offers a user-friendly interface and powerful features like live chat, email integration, and automated responses, making it a favorite among small businesses.
- **LivePerson:** Specializes in conversational commerce, enabling businesses to have meaningful interactions with customers across various messaging platforms.
- **Zendesk Chat:** Part of the Zendesk suite, it offers a comprehensive solution for customer service, including AI-driven chat capabilities.

How to Integrate Your Chatbot with ChatGPT

Integrating your chatbot with ChatGPT can significantly enhance its conversational abilities, making interactions more natural and engaging.

1. **Choose a Platform:** Select a chatbot platform that supports AI integrations, such as Intercom, Drift, or Tidio.
2. **API Access:** Obtain API access to ChatGPT from OpenAI. You will need an API key to connect your chatbot to ChatGPT.
3. **Integration Process:** Follow the integration guidelines provided by your chatbot platform. This typically involves configuring API endpoints and setting up webhook events.
4. **Training and Customization:** Customize the responses of your ChatGPT integration to match your brand's voice and tone. Provide it with relevant information about your products, services, and FAQs to improve its accuracy.
5. **Testing:** Test the integration thoroughly to ensure that the chatbot is providing accurate and helpful responses. Make adjustments as needed based on user feedback and interaction data.

Little-Known Tips and Tricks

Here are some lesser-known tips and tricks to get the most out of your AI chatbot:

- **Personalization:** Train your chatbot to recognize returning visitors and personalize the conversation based on previous interactions. This can make users feel valued and increase their engagement.
- **Use Rich Media:** Enhance your chatbot's responses with rich media elements like images, GIFs, and videos. This can make the interaction more engaging and informative.
- **A/B Testing:** Regularly conduct A/B testing on your chatbot scripts to identify the most effective responses. This helps in continuously improving the chatbot's performance.
- **Feedback Loop:** Implement a feedback mechanism where users can rate their chatbot experience. Use this data to make improvements and address any issues.
- **Multi-Language Support:** Expand your chatbot's reach by incorporating multi-language support. This is especially useful if you have a global audience.

By deploying an AI chatbot on your website, you not only enhance user engagement but also streamline your customer support processes. Embrace this technology to stay ahead of the competition and provide an exceptional experience for your visitors.

DAY 23: AI-DRIVEN SEO OPTIMIZATION

In simplest terms, SEO, or Search Engine Optimization, is the practice of improving your website so it ranks higher on Google and other search engines. The better visibility your pages have in search results, the more likely you are to garner attention and attract prospective and existing customers to your business. It involves on-page SEO (content and HTML source code) and off-page SEO (backlinks and other external signals).

In today's competitive digital landscape, leveraging AI-driven SEO optimization is crucial for businesses looking to enhance their online presence. AI tools can continuously analyze and optimize your website's SEO, driving more organic traffic and capturing leads through improved search rankings. Let's dive into how you can use AI to boost your SEO efforts.

How to Rank at the Top of Google

Ranking at the top of Google involves a combination of traditional SEO practices and leveraging the latest AI tools. Here are some key strategies that are working right now, but keep in mind that the algorithm is constantly changing:

- **High-Quality Content**: Google prioritizes content that is valuable, relevant, and well-written. Ensure your content answers the queries of your target audience comprehensively.
- **User Experience (UX)**: Google's algorithms increasingly focus on user experience. Your site should be mobile-friendly, fast-loading, and easy to navigate.
- **E-A-T (Expertise, Authoritativeness, Trustworthiness)**: Google values content from experts and trusted sources. Building a robust backlink profile from reputable sites and maintaining accurate, informative content is essential.
- **Voice Search Optimization**: With the rise of voice search, optimize for conversational queries and long-tail keywords.
- **Technical SEO**: Ensure your site is technically sound, with proper indexing, XML sitemaps, and structured data.

Which AI Tools Are the Best for SEO Right Now?

- **SurferSEO**: This tool helps you optimize your content by providing data-driven recommendations based on the top-performing pages for your keywords.
- **Ahrefs**: Known for its robust backlink analysis, keyword research, and competitive analysis.
- **SEMrush**: Offers comprehensive SEO tools, including keyword research, site audit, and position tracking.
- **Frase**: Uses AI to help you create content that answers user queries better than your competitors.
- **BrightEdge**: Provides data-driven SEO recommendations and tracks your performance against competitors.

How to Use ChatGPT to Help with SEO

ChatGPT can be a powerful tool for SEO in several ways:

- **Content Creation**: Use ChatGPT to generate high-quality, relevant content tailored to your audience's needs. It can help with blog posts, product descriptions, and more.
- **Keyword Research**: Ask ChatGPT to suggest long-tail keywords or phrases that users might search for related to your topic.
- **SEO Audit**: Use ChatGPT to generate a checklist of SEO best practices and ensure your website follows them.
- **Meta Descriptions and Titles**: Generate compelling meta descriptions and title tags to improve click-through rates.
- **Content Ideas**: Get ideas for new content topics by asking ChatGPT about trending topics or common questions in your industry.

The Future of SEO in an AI World

The landscape of SEO is evolving with AI, and the future holds exciting possibilities:

- **Conversational AI**: More people will use conversational AI like ChatGPT instead of traditional search engines. Optimize your content for AI-driven platforms by focusing on natural language and conversational keywords.

- **AI-Generated Content**: AI tools will increasingly assist in content creation, enabling faster and more efficient production of high-quality content.
- **Personalized Search**: AI will allow for more personalized search results, meaning your SEO strategy must consider individual user intent and preferences.

As more users turn to conversational AI for their information needs, ensure your website is designed to cater to these platforms. This means creating content that is easily digestible by AI, using structured data, and ensuring your site's information is accurate and authoritative.

Generative Engine Optimization (GEO) is an emerging approach that combines traditional SEO with AI-driven techniques to enhance search visibility. Unlike standard SEO, which focuses on keywords and backlinks, GEO optimizes content for AI-based search engines like ChatGPT. This involves creating high-quality, authoritative content that answers user queries directly and comprehensively. Early adoption of GEO strategies can significantly improve organic growth and engagement, much like the early days of platforms like Facebook.

Here's a quick hack to optimize your site for generative engines: begin your content with a question. Try this prompt:

> Act as an expert copywriter with specialty in optimizing content for SEO.
> What are the most frequently asked questions about [your industry, product, or service]?
>
> List the questions, then help me write blog posts to answer each question.

By implementing AI-driven SEO strategies, you can stay ahead of the curve and maintain a strong online presence in an ever-evolving digital world.

DAY 24: ASK YOUR EXISTING CLIENTS FOR REFERRALS

Referrals are a powerful tool in growing your business. They come with a built-in trust factor because they are recommended by someone the potential customer already knows and trusts. Here's how you can leverage your existing clients to generate referrals, and how to use AI to streamline and enhance this process.

Referrals are the lifeblood of any business, and here's why:

- Higher Conversion Rates: Referred customers are more likely to convert into paying clients because they come with a pre-established trust.
- Cost-Effective: It costs less to acquire new clients through referrals compared to traditional advertising methods.
- Increased Loyalty: Customers who come through referrals tend to be more loyal and have a higher lifetime value.
- Brand Advocacy: Satisfied customers who refer others become advocates for your brand, helping to spread positive word-of-mouth.

How to Use ChatGPT for This

ChatGPT can be a game-changer in crafting personalized and effective referral requests. Here's how to utilize it:

- Drafting Referral Requests:
    - Email: Use ChatGPT to draft personalized referral request emails. For example:

        "Hey [Client's Name], we hope you're enjoying [Product/Service]. We strive to provide excellent service and would appreciate it if you could recommend us to your friends or colleagues. As a thank you, you'll receive [Incentive] for every successful referral!"

    - Message Scripts: Create scripts for phone calls or text messages asking for referrals.

- Social Media Posts: Generate posts that encourage your followers to refer friends to your business.
- Follow-Up Messages:
    - Reminder Emails: Set up automated reminders for clients who haven't responded to the initial referral request.
    - Thank You Messages: Use ChatGPT to draft thank you messages for clients who have referred others, showing appreciation and fostering loyalty.

The Top Tools That Can Help

Several tools can assist you in managing and optimizing your referral program. Here are a few:

- **ReferralCandy:** Automates your referral program and integrates with your e-commerce platform.
- **Referral Rock:** Offers robust features for managing and tracking referrals, and integrates with various marketing tools.
- **Yotpo:** Provides tools for managing customer reviews and referrals, enhancing your brand's credibility.
- **Influitive:** Helps create engaging referral campaigns and tracks performance metrics.

How to Automate (or Semi-Automate) This Process

- **Email Automation:**
    - Use platforms like Mailchimp or ActiveCampaign to send automated referral request emails to your clients.
    - Set up drip campaigns that follow up with clients who haven't responded to initial requests.
- **CRM Integration:**
    - Integrate your referral program with your CRM system to track and manage referrals efficiently.
    - Use CRM tools like HubSpot or Salesforce to automate follow-ups and thank you messages.

- **Incentive Management:**
  - Automate the distribution of referral incentives using platforms like Gusto or Stripe.
  - Track referrals and rewards to ensure timely and accurate incentive distribution.

Little-Known Tips and Tricks

- **Timing is Key:** Ask for referrals when your clients are happiest, such as right after a successful project or positive feedback. At my company Zeus' Closet, when clients pick up their customized apparel and are in a high emotional state, we immediately ask if they would leave us a 5-star review and we text them the direct link.
- **Personalize Requests:** Tailor your referral requests to each client, mentioning specific interactions or projects to make the request more personal.
- **Leverage Social Proof:** Share success stories or testimonials from other clients to encourage referrals.
- **Referral Forms:** Create simple, easy-to-fill referral forms on your website or via email to reduce friction.
- **Offer Exclusive Incentives:** Provide unique incentives that are appealing to your clients, such as discounts on future purchases or exclusive access to new products. Be discreet when doing this though, as it could violate the terms of some review sites.

By leveraging the power of AI and automation, you can create a seamless and effective referral program that taps into the potential of your existing client base. This approach not only enhances client satisfaction but also drives new business growth efficiently.

DAY 25: NEWSJACKING

Newsjacking is the art of leveraging breaking news to amplify your brand's message. Coined by David Meerman Scott in his book *Newsjacking: How to Inject Your Ideas into a Breaking News Story and Generate Tons of Media Coverage*, the concept revolves around capitalizing on trending news to gain media attention and boost your marketing efforts.

Scott observed that traditional PR strategies were slow and often missed the opportunity to capitalize on rapidly evolving news stories. Newsjacking, he argued, allows brands to insert themselves into ongoing conversations, making their messages more relevant and timely.

For instance, Oreo's famous tweet during the 2013 Super Bowl blackout— "You can still dunk in the dark"—is a classic example of newsjacking. Oreo seized the moment, creating a viral sensation and demonstrating the power of real-time marketing.

Why it Works

Newsjacking works because it taps into the natural human curiosity and the need to stay updated with current events. When a story breaks, people are searching for information and updates. By aligning your brand with a trending topic, you capture this attention and redirect it towards your offerings.

Consider how Tesla often garners media attention by aligning its product launches with significant news events or trends in the renewable energy sector. This strategic alignment not only enhances brand visibility but also positions Tesla as a thought leader in the industry.

How to Do It Using AI to Help

Using AI for newsjacking can streamline the process and enhance effectiveness. Here's a step-by-step guide:

Monitor Trending Topics: Use AI tools like Google Trends, BuzzSumo, or TrendHunter to keep an eye on emerging news stories relevant to your industry.

Generate Ideas: Leverage AI to brainstorm content ideas. For example, use ChatGPT with the following prompt:

> Act as a ghostwriter who specializes in newsjacking. Search the web for the top news headlines from the past 2 days, and give me three content ideas that weave my offerings with the most current events.

This prompt helps you quickly generate relevant content ideas aligned with trending topics.

Then continue the conversation with ChatGPT and ask it to draft articles, social media posts, or press releases that integrate the trending news with your brand message. Refer to some of the earlier prompts we used.

Distribute quickly. Leverage social media scheduling tools like Hootsuite or Buffer to distribute your content promptly. The speed of your response is crucial in newsjacking.

Newsjacking is a powerful tool in the modern marketer's arsenal.

By leveraging AI to monitor trends, generate content, and distribute it quickly, you can effectively insert your brand into the heart of ongoing conversations, driving visibility and engagement.

DAY 26: BE A GUEST ON PODCASTS

Being a guest on podcasts involves participating in other people's podcast shows to share your expertise, promote your brand, or discuss topics related to your field. This approach allows you to tap into existing audiences who are already interested in your niche, providing a platform to expand your reach and establish credibility.

The concept of guest appearances in the media isn't new. It dates back to traditional media, where experts and celebrities would appear on radio shows, television talk shows, and news segments. As podcasts emerged in the early 2000s, they quickly became a modern extension of this practice, offering more specialized and accessible platforms for thought leaders and industry experts to share their insights.

Why it Works

Guest appearances on podcasts work because they provide a targeted audience that is already engaged and interested in your topic. Podcasts have a loyal following, with listeners who trust the host's recommendations and are open to new ideas and voices. By appearing on these shows, you leverage the host's credibility and rapport with their audience, creating a powerful marketing channel.

When I launched *Wealth Beyond Money*, I went on dozens of podcasts to promote the book. This strategy allowed me to reach thousands of potential readers who were already interested in personal development, mental health, and business growth. The intimate and conversational nature of podcasts helped me build a personal connection with listeners, which translated into a bestselling book and increased engagement with my brand.

How to Do it Using AI to Help

Leveraging AI can streamline the process of becoming a guest on podcasts. Here are a few ways AI can assist:

1. **Research and Outreach:**

Use AI tools like ChatGPT to draft personalized outreach emails to podcast hosts. AI-powered platforms can help identify the most relevant podcasts based on your niche and audience. Here's a prompt to try:

> Based on what you know about me, search the web for good podcasts that I could be a guest on to promote _____. Include contact information to make it easy for me to reach out to the host.

2. **Preparation:**

ChatGPT can help generate potential interview questions and talking points based on the podcast's focus. AI can analyze past episodes to understand the host's style and the topics covered, allowing you to tailor your message effectively. Try this prompt next:

> Help me prepare for the podcasts. What are some questions the host might ask me? Help me think through answers to those questions.

3. **Promotion:**

After your appearance, AI can help create social media posts, summaries, and even audiograms to promote the episode across different platforms. Refer to the earlier tools on how to do this. Leveraging podcast appearances as long-form video to chop up into shorter clips is a great strategy.

Ways to Automate (or Semi-Automate) This Process

Automation can significantly reduce the time and effort required to secure podcast appearances. Here are some steps you can automate:

1. **Identifying Podcasts:**

   Use tools like Podchaser, Listen Notes, or PodMatch to automate the search for podcasts in your niche.

2. **Email Outreach:**

   Platforms like Mailshake or Lemlist can automate email campaigns to podcast hosts, including follow-up sequences.

3. **Content Creation:**

   AI tools like Jasper can help automate the creation of promotional content post-interview. Or use your personally-trained ChatGPT or Claude. Whichever one gives you the best results.

Some of the Top Tools for Podcast Guesting

- **Podchaser:** A comprehensive podcast database for discovering relevant shows and connecting with hosts.
- **Listen Notes:** A search engine for podcasts that helps identify opportunities based on your niche.
- **PodMatch:** An AI-driven platform that matches podcast guests with hosts, simplifying the process of finding relevant shows.

Pro Tips and Tricks

- Leverage LinkedIn: Use LinkedIn to connect with podcast hosts and engage with their content before reaching out.
- Create a Media Kit: Prepare a media kit with your bio, headshot, book information, and key talking points to make it easy for hosts to feature you. Canva is a great tool for this.

- Have answers prepared: The podcast is your chance to get your message across. Don't stumble over your words. Prepare in advance and keep a document of possible interview answers ready. You don't want to read your answers like a script, but you want to be extremely familiar with the content so it rolls off the tongue. Always focus on how you can provide value to the host's audience rather than just promoting your product or service.
- Repurpose Content: Turn podcast interviews into blog posts, social media snippets, and newsletters to maximize exposure.
- Follow Up: Send a thank-you note and share the episode with your audience, strengthening your relationship with the host.

Action Item

To get started, use this prompt with ChatGPT to draft your initial outreach email:

> Write a compelling email to a podcast host, introducing myself [who you are and your expertise] expressing my interest in being a guest on their show to discuss actionable strategies for [topic], and highlighting how my insights can provide value to their listeners.

DAY 27: TRADE SHOWS

Almost every industry has a trade show. Trade shows are events where companies and professionals within a specific industry gather to showcase their products, services, and innovations. These events provide an opportunity for businesses to connect with potential clients, partners, and industry leaders. Trade shows are also a platform for networking, learning about industry trends, and gaining competitive insights.

Trade shows have a long history dating back to medieval Europe, where merchants would gather at fairs to display their goods. Over time, these gatherings evolved into more organized events focused on specific industries. The modern trade show as we know it began to take shape in the 20th century, with the advent of large-scale exhibitions and conventions. Today, trade shows are integral to business strategies across various sectors, providing a venue for marketing, networking, and business development.

Trade shows are effective for several reasons:

- **Direct Interaction**: They provide a platform for face-to-face interaction with potential clients and partners.
- **Exposure**: Companies can showcase their products to a large, targeted audience.
- **Networking**: Trade shows are prime opportunities to connect with industry peers and influencers.
- **Market Insights**: Attendees gain valuable insights into industry trends and competitor strategies.
- **Lead Generation**: They are excellent for generating high-quality leads and building brand awareness.

Little-Known Tips and Tricks

1. **Leverage Social Media**: Use social media to create buzz before, during, and after the event. Live-tweeting, posting stories, and engaging with attendees online can enhance your visibility.
2. **Host a Side Event**: Organize a small gathering or workshop during the trade show to connect more intimately with key prospects. Make sure to document this well for social media content too.
3. **Use QR Codes**: Simplify the lead capture process by using QR codes that attendees can scan to get more information about your products or services.
4. **Follow-Up Strategy**: Have a clear follow-up strategy in place. Use automated email sequences to nurture leads post-event.

Action Items

1. **Identify Key Events**: Use the following prompt to generate a list of relevant trade shows and industry events:

> Act as my Chief Marketing Officer. Based on what you know about me, list relevant trade shows and industry events I could attend, and tell me the best ways to leverage my attendance/participation at each event. Include links for each event.

2. **Plan Your Participation**: Research each event and determine how best to participate. Consider speaking opportunities, sponsorships, or exhibiting.
3. **Prepare Marketing Materials**: Create and personalize marketing materials tailored to each trade show audience.
4. **Automate Follow-Ups**: Set up automated follow-up emails and CRM integration to capture and nurture leads.

By strategically leveraging trade shows and incorporating AI tools, you can maximize your impact, generate high-quality leads, and strengthen your industry presence.

DAY 28: WRITE CASE STUDIES

Case studies are in-depth examinations of specific situations, challenges, or projects, highlighting how particular solutions or strategies led to successful outcomes. They serve as powerful tools to demonstrate expertise, showcase results, and build credibility. Case studies typically include the background of the challenge, the approach taken, and the results achieved.

Case studies work because they provide real-world examples of how theories or strategies can be applied effectively. They offer proof of success, helping to build trust and credibility with potential clients or customers. Additionally, they serve as educational tools, illustrating best practices and lessons learned.

How to Write Effective Case Studies

To write an effective case study, follow these steps:

1. Identify the Subject: Choose a project, client, or situation that highlights your strengths and showcases your success.
2. Gather Data: Collect all relevant information, including project details, client feedback, and measurable results.
3. Structure the Story: Organize the case study into a clear narrative, with sections for the challenge, approach, and results (CAR)
4. Include Visuals: Use photos, charts, and graphs to make the case study visually appealing and easier to understand.
5. Highlight Key Insights: Focus on the lessons learned and the impact of the project.

At Stuff4GREEKS, we run an annual Jacket of the Year contest, where each entry serves as a case study. This contest not only highlights our custom jacket designs but also tells the story of transformation for each participating group. Here's a detailed look at how we turn each entry into a compelling case study:

1. Challenge: Each group faces the challenge of creating a unique, memorable jacket design that represents their identity and values.

2. Approach: We work closely with the group to understand their vision, providing expert design advice and multiple iterations to perfect the design.
3. Results: The final jackets are not only visually stunning but also hold significant sentimental value. Each story is accompanied by before-and-after photos, showcasing the transformation.

For poetic purposes, we call our sections "The Dream," "The Design," and "The Delivery" but they mean the same thing as above. You can see examples at https://s4g.com/joty

Leveraging AI to Create Case Studies

Writing case studies is an essential skill for showcasing your successes and demonstrating your expertise. By following a structured approach and leveraging AI tools, you can create compelling narratives that resonate with your audience and highlight the impact of your work.

Action Items

1. Try this prompt:

> Act as a professional case study writer. Based on the following data and client feedback, draft a compelling case study highlighting the challenge, approach, and results:
> - Client:
> - Challenge:
> - Our approach:
> - Results:

2. Include Visuals: Enhance your case study with photos and other visuals to illustrate the transformation.

By consistently creating and sharing detailed case studies, you can build trust, establish credibility, and showcase the real-world impact of your solutions.

DAY 29: BUY LEADS

Lead buying is a marketing strategy where businesses purchase contact information of potential customers, or "leads," from third-party vendors. These leads are typically categorized based on demographics, behaviors, and other criteria to ensure they match the business's target audience. This approach can significantly boost marketing efforts by providing a steady stream of potential customers ready for engagement.

Buying leads can be advantageous for several reasons:

- Speed: It accelerates the process of finding potential customers, saving time on prospecting.
- Targeted Marketing: Purchased leads are often pre-qualified and categorized, ensuring that your marketing efforts are directed at a relevant audience.
- Scalability: It allows for scaling up your marketing efforts quickly, supporting business growth.
- Cost-Effectiveness: It can be more cost-effective than traditional lead generation methods, such as outbound sales and advertising.

How to Buy Leads Effectively

To effectively buy leads, follow these steps:

1. Identify Your Target Audience: Understand the specific demographics, behaviors, and needs of your ideal customers.
2. Choose a Reliable Vendor: Select a reputable lead provider with a track record of delivering high-quality leads.
3. Specify Criteria: Clearly define the criteria for your leads, including industry, job title, location, and other relevant factors.
4. Test and Evaluate: Start with a small batch of leads to evaluate their quality and conversion potential before making larger purchases.
5. Integrate with CRM: Use a Customer Relationship Management (CRM) system to integrate and manage your purchased leads effectively.

AI can enhance the lead buying process in several ways:

- Predictive Analytics: AI tools can analyze large datasets to predict which leads are most likely to convert.
- Lead Scoring: AI can score leads based on their potential value, helping prioritize follow-up actions.
- Personalized Marketing: AI-driven insights can help tailor marketing messages to individual leads, increasing engagement and conversion rates.

Prompt Example for AI-Enhanced Lead Buying:

> Act as a lead generation expert. Based on my business model, identify the best criteria for purchasing leads and suggest a list of reputable vendors. Additionally, recommend AI tools that can help optimize the lead buying process.

Buying leads is a powerful strategy to quickly and effectively expand your customer base. By identifying the right target audience, choosing reliable vendors, and leveraging AI tools, you can optimize your lead buying process and drive significant business growth.

DAY 30: COLD CALLING SCRIPTS

One of my first jobs in college was as a telemarketer at a company called DialAmerica. The computer would auto-dial people and I would read a script. That job had high turnover because we dealt with constant rejection, but I eventually internalized a deep understanding of the product's benefits for the person on the other end of the phone, and I got pretty good at it.

Cold calling is the practice of reaching out to potential customers who have had no prior contact with your business. It involves making unsolicited calls to individuals or businesses with the aim of generating new leads, setting appointments, or making sales. Despite its challenging nature, cold calling remains a powerful tool for direct engagement and lead generation when executed correctly.

Cold calling works because it allows for direct, personal interaction with potential customers. It provides immediate feedback and the opportunity to address questions and objections in real time. When done well, cold calling can build relationships, uncover new opportunities, and drive sales.

To create an effective cold calling script, follow these steps:

1. Research Your Audience: Understand the needs, pain points, and goals of your target audience. We covered this in an earlier step.
2. Craft a Strong Opening: Start with a compelling introduction that grabs attention and establishes credibility.
3. Present a Clear Value Proposition: Clearly explain the benefits of your product or service and how it addresses the prospect's needs.
4. Ask Open-Ended Questions: Open-ended questions typically start with "what" or "how," compared to closed-ended questions which can be answered "yes" or "no." Engage the prospect by asking questions that encourage dialogue and uncover valuable information.
5. Handle Objections Gracefully: Prepare for common objections and develop strategies to address them effectively.

6. Close with a Call to Action: End the call with a clear next step, such as scheduling a meeting or sending additional information.

Leveraging AI to Enhance Cold Calling

AI can enhance cold calling by:

1. Prospect Research: AI tools can gather and analyze data to help identify and prioritize high-potential leads.
2. Script Optimization: AI can analyze call recordings to identify effective phrasing and suggest improvements.
3. Automated Follow-Up: AI-driven CRM systems can automate follow-up emails and reminders to ensure timely engagement with leads.

Try this prompt in the same chat from earlier:

> Act as an expert sales consultant. Based on my target audience of [your target market] and their pain points, draft a cold calling script for [your company] that includes a strong opening, a clear value proposition, engaging open-ended questions, and strategies for handling common objections. Emphasize the benefits immediately and often.

Action Items

1. Research Your Target Audience: Use ChatGPT to gather insights on their needs and pain points.
2. Draft Your Script: Use the prompt above to create a compelling cold calling script.
3. Practice and Refine: Rehearse your script and adjust based on feedback and performance.
4. Leverage AI Tools: Use AI for prospect research, script optimization, and automated follow-ups.
5. Lean Into Rejection: The more "no's" you get, the more you learn.

DAY 31: BILLBOARDS

Billboards may seem old school, but done right, and for the right type of businesses, they can work. Billboards are designed to catch the attention of motorists and pedestrians, offering a bold, concise message that sticks in the viewer's mind.

The concept of billboard advertising dates back to ancient civilizations. For example, ancient Egyptians used tall stone obelisks to publicize laws and decrees. However, the modern billboard, as we know it, emerged in the late 19th century, coinciding with the rise of the automobile. The need to capture the attention of drivers led to the creation of large, eye-catching advertisements positioned along popular roadways.

Despite the rise of digital marketing, billboards remain an effective advertising medium. Their large size and strategic placement make them hard to ignore. According to a Nielsen study, 66% of people surveyed found billboards to be a good way to learn about new businesses and events.[5] Additionally, 40% visited an advertiser because of a billboard, and 24% made a purchase because of one. This demonstrates billboards' strong ability to drive consumer action.[6]

At my company Zeus' Closet, we decided to test the power of billboards by launching a campaign with the tagline: "Like a tattoo shop for clothes." This seven-word message effectively communicated our brand's unique value proposition and was easy to remember. The campaign significantly boosted our brand recognition and foot traffic, underscoring the effectiveness of concise messaging in billboard advertising.

AI can significantly enhance the effectiveness of billboard campaigns in several ways:

1. Data Analysis: AI can analyze traffic patterns and demographic data to identify optimal billboard locations.
2. Design Optimization: AI tools can analyze successful ads to recommend the best designs, including layout, colors, and fonts.

---

[5] "Brand New Nielsen Study On Billboard Advertising: High Awareness, Engagement, And Brand Action," *View Chicago*, May 31, 2019, https://www.vcoutdoor.com/blog/2019/5/29/brand-new-nielsen-study-on-billboard-advertising-high-awareness-engagement-and-brand-action

[6] Wade Swormstedt, "2017 Nielsen Study Examines Billboard Effectiveness," FASI, https://www.signreference.org/2018/01/19/2017-nielsen-study-examines-billboard-effectiveness/

3. Performance Tracking: AI can track billboard effectiveness by analyzing social media mentions, website traffic, and sales data.

Ways to Automate (or Semi-Automate) This Process

1. Location Selection: Use AI-powered platforms like Geopath to choose the best locations based on traffic and demographic data.
2. Ad Creation: Tools like Canva and Adobe Spark, enhanced with AI, streamline the design process.
3. Monitoring and Adjusting: Platforms like AdQuick allow you to monitor the performance of your billboards in real-time and make necessary adjustments.

Little Known Tips and Tricks

- Keep It Short: Aim for seven words or less, as exemplified by our "Like a tattoo shop for clothes" tagline.
- Use Bold, Readable Fonts: Ensure your message is legible from a distance.
- High Contrast Colors: Use colors that stand out and are easily visible.
- Strategic Placement: Position billboards where your target audience is most likely to see them.

Action Item

To get started with creating a compelling billboard campaign using AI, try this prompt in ChatGPT:

> Act as an expert marketer, and help me create a billboard campaign for my business [your business name]. The campaign should include a catchy seven-word tagline, design suggestions, and the best locations for placement based on [your target market].
>
> Additionally, suggest tools for tracking and optimizing the campaign's performance. Conclude by creating some image concepts for the billboard.

Yes, if you have the paid version of ChatGPT, it will even use DALL-E to create some mockup images of a billboard for you!

By leveraging these insights and tools, you can create effective billboard campaigns that drive significant consumer engagement and business growth.

DAY 32: DIRECT MAIL

Direct mail is a form of advertising where businesses send physical promotional materials to potential customers via postal services. Unlike digital marketing, which relies on email and online ads, direct mail involves sending postcards, letters, brochures, or catalogs directly to consumers' mailboxes. This tangible approach can create a more personal and impactful connection with the audience.

Direct mail has a rich history dating back to the late 19th and early 20th centuries. Companies began leveraging the postal system to reach customers with promotional offers and product information. One of the earliest and most notable successes in direct mail marketing was the Sears, Roebuck and Co. catalog, first published in 1888. It allowed consumers, especially in rural areas, to purchase goods that were previously inaccessible, transforming Sears into a household name and a retail giant.

Direct mail works because it engages multiple senses, making it more memorable than digital ads that can be easily ignored or forgotten. A physical piece of mail can be touched, held, and even smelled, creating a multi-sensory experience that enhances recall. Additionally, direct mail often feels more personal, as it arrives directly at the recipient's home and can be customized with their name and specific offers tailored to their interests.

For instance, I pass by a certain restaurant every day but always forget it's there because it sits back from the street. One day, we received a postcard with three coupons for that very restaurant. The tangible reminder and the value of the coupons prompted us to try the restaurant. This example shows how direct mail can effectively bring a business top-of-mind and drive foot traffic.

AI can significantly enhance the effectiveness of direct mail campaigns. Here's how:

- Targeting and Segmentation: AI algorithms can analyze large datasets to identify the most promising customer segments based on purchasing behavior, demographics, and interests. This ensures that your direct mail reaches the right audience.

- Personalization: AI can customize the content of your direct mail pieces, making them more relevant to each recipient. Personalized offers, product recommendations, and even personalized imagery can increase engagement and conversion rates.
- A/B Testing and Optimization: AI can help you conduct A/B tests to determine which direct mail pieces are most effective. By analyzing response rates and other metrics, AI can suggest optimizations to improve future campaigns.
- Predictive Analytics: AI can predict which customers are most likely to respond to direct mail based on their past interactions and behaviors. This helps in allocating resources more efficiently and improving ROI.

Ways to Automate (or Semi-Automate) This Process

- Mail Merge Tools: Use software that can merge your customer data with direct mail templates to create personalized mail pieces at scale. Tools like Microsoft Word's Mail Merge or specialized software like Lob can automate this process.
- Customer Relationship Management (CRM) Systems: Integrate your CRM with direct mail services to automatically trigger mail sends based on customer actions or lifecycle stages. For instance, a CRM like Salesforce can work with direct mail platforms to send postcards to new customers or those who haven't visited in a while.
- AI-Powered Design Tools: Use AI design tools to create visually appealing direct mail pieces quickly. Tools like Canva or Adobe Spark use AI to suggest design elements and layouts based on your input.

Some of the Top Tools for Direct Mail

- Lob: A platform that automates direct mail campaigns with powerful APIs for creating, sending, and tracking mail.
- Sendoso: Combines digital and physical marketing by automating the sending of personalized gifts and direct mail.
- PFL: Integrates with CRM and marketing automation platforms to trigger direct mail sends based on digital behavior.

- Postable: Simplifies the process of creating and sending personalized cards and notes.
- Click2Mail: Provides an easy-to-use online platform for designing and sending direct mail.

Little Known Tips and Tricks

- Handwritten Notes: Consider adding a handwritten element to your direct mail pieces. Even a short, personalized note can significantly increase engagement.
- Include a QR Code: There needs to be a compelling reason for them to scan the code and go to the web page that is your link. Consider offering a coupon or resource guide. When they get to the web page, capture their email address in exchange for the goodie(s) you promised.
- Augmented Reality (AR): Incorporate AR elements that recipients can scan with their smartphones to bring your mail piece to life with interactive content.
- Eco-Friendly Options: Use recycled materials and eco-friendly inks to appeal to environmentally conscious consumers.
- Follow-Up with Digital: Send a follow-up email or digital message referencing the direct mail piece to reinforce the message and encourage action.

Action Item

Here's a ChatGPT prompt to help you brainstorm your next direct mail campaign:

> Create a direct mail campaign for my business [your business name] that targets [your target market for this campaign]. The campaign should include a personalized message, a special offer, and a visually appealing design. Create example images. Include ideas for follow-up actions to maximize engagement and conversions.

Direct mail remains a powerful tool in the marketer's arsenal, especially when enhanced with modern AI technologies. By understanding the principles of effective direct mail, leveraging the best tools, and employing innovative strategies, you can create compelling campaigns that drive significant business growth. Remember, the key to success is personalization, targeting, and a seamless integration of your physical and digital marketing efforts.

DAY 33: INFLUENCER MARKETING

The concept of influencer marketing isn't new. It dates back to the early 20th century when companies began using celebrities to endorse their products. However, the rise of social media platforms like Instagram, YouTube, and TikTok has democratized influence, allowing everyday individuals to amass large followings and become powerful brand advocates. This shift has enabled businesses to reach more targeted and engaged audiences through influencers who resonate with specific demographics.

Influencer marketing works because people trust recommendations from individuals they follow and admire more than traditional ads. Influencers have built strong relationships with their audience, and their endorsements can significantly sway purchasing decisions. According to a 2023 study by Edelman, 68% of consumers trust recommendations from people they know,[7] and 63% have noticed an increase in sponsored content from influencers on social media in the past year.[8] This trend underscores the importance of authenticity and the persuasive power of influencer endorsements in modern marketing.

A YouTuber named Corey Jones once reached out to us. We made a custom jacket for him in exchange for an unboxing video and a video tour of our headquarters. It was a win-win. Corey has a unique jacket and lots of shoppers who told us they heard about us from Corey's video. The videos are still up, and they continue to get lots of views (search YouTube for "Corey Jones stuff4GREEKS").

Some of the Top Tools for Influencer Marketing

- AspireIQ: Known for its robust influencer discovery and relationship management features.
- Influencity: Offers detailed analytics and audience insights to find the best influencers for your brand.
- Traackr: Provides comprehensive influencer data and performance metrics.

---

[7] "The Collapse of the Purchase Funnel," *Edelman*, 2023, https://www.edelman.com/trust/2023/trust-barometer/special-report-brand-trust

[8] "Consumers Continue To Seek Influencers Who Keep It Real," *Matter*, Feb. 22, 2023, https://www.matternow.com/blog/consumers-seek-influencers-who-keep-it-real/

Little Known Tips and Tricks

- Leverage Nano Influencers: Nano influencers (those with 1,000 to 10,000 followers) are often more affordable and willing to barter. They also tend to have higher engagement rates and a more loyal following. Coery Jones, who we mentioned earlier, had about 10,000 subscribers when we first worked with him, but now he has over 100,000 subscribers.
- Long-Term Partnerships: Building long-term relationships with influencers can lead to more authentic and consistent promotion.
- Exclusive Offers: Give influencers exclusive discounts or early access to products. This not only makes them feel valued but also encourages their followers to act quickly.
- Content Repurposing: Repurpose influencer content across your own social media channels, website, and marketing materials to maximize its impact.
- Engage with Their Audience: Don't just focus on the influencer; engage with their followers through comments and direct messages to build a broader community.

Action Item: Use AI to Help

AI can streamline your influencer marketing efforts, making it easier to find the right influencers and measure the success of your campaigns. Try this prompt:

> I want to launch an influencer marketing campaign for [your company]. Outline a plan for me. Who should I approach? What should I say? What can I offer them? What else do I need to know?

DAY 34: BOOST ORGANIC POSTS INTO PAID ADS

Boosting organic posts into paid ads involves taking content that has already been published on your social media platforms and promoting it through paid advertising. This technique leverages the natural engagement and success of your organic posts, providing a cost-effective and strategic approach to amplifying your reach and driving more substantial results. By identifying posts that are already performing well, you can ensure that your ad spend is directed towards content that resonates with your audience.

The idea of boosting organic posts emerged with the advent of social media advertising. As platforms like Facebook, Instagram, and Google evolved, they offered businesses the ability to promote content directly. Initially, companies would create separate ads, but marketers soon realized that promoting organic posts that already had engagement could yield better results. This approach capitalizes on social proof, as users are more likely to engage with content that others have already interacted with positively.

Boosting organic posts works because it combines the strengths of both organic and paid strategies. Organic posts that perform well do so because they resonate with your audience, whether through compelling visuals, engaging copy, or timely relevance. When you boost these posts, you amplify their reach to a broader audience who are likely to engage similarly. This method ensures that your ad budget is used efficiently, targeting content with a proven track record. Additionally, the engagement metrics from organic performance provide valuable insights into what your audience prefers, enabling more informed and effective ad campaigns.

How to Do It Using AI to Help

- Identify High-Performing Posts: Use AI-driven tools to analyze the engagement metrics of your organic posts. Tools like Sprout Social or Hootsuite can provide detailed insights into likes, shares, comments, and overall engagement.

- Predictive Analytics: AI can help predict which posts are likely to perform well if boosted. Platforms like Cortex use machine learning to analyze past performance and suggest content with high potential.
- Audience Segmentation: AI can refine your audience targeting. Tools like Facebook's Lookalike Audiences use machine learning to find new potential customers similar to your existing followers.
- Ad Creative Optimization: AI-powered tools like AdCreative.ai can help optimize your ad creative by suggesting the best images, headlines, and descriptions based on historical data and trends.

Ways to Automate (or Semi-Automate) This Process

- Automated Boosting: Platforms like Hootsuite and Buffer offer features to automatically boost posts that reach a certain engagement threshold.
- Ad Scheduling: Use tools like AdEspresso to schedule your boosted posts for optimal times based on when your audience is most active.
- Budget Management: Automate budget allocation with AI tools that adjust spending based on real-time performance data, ensuring you get the best ROI.
- Performance Tracking: Utilize AI analytics tools to continuously monitor the performance of your boosted posts, making real-time adjustments to maximize effectiveness.

Little Known Tips and Tricks

- Test Different Formats: Experiment with boosting different types of content – videos, carousel posts, and static images – to see what works best.
- Leverage User-Generated Content: Boost posts featuring user-generated content to build trust and authenticity.
- Utilize Seasonal Trends: Boost posts that align with current trends or seasons to increase relevance.
- Optimize for Conversions: Use analytics to track which boosted posts drive the most conversions and adjust your strategy accordingly.
- Retargeting: Create retargeting campaigns for users who engage with your boosted posts to move them further down the sales funnel. We will cover this more later.

DAY 35: WRITE A BOOK

I know... writing a book might feel more overwhelming than anything else we've covered so far. For me, writing a book is like running a marathon. I get excited about the idea of doing it–it sucks *while* I'm doing it–but when I'm done, I'm glad I did it.

The analogy of a marathon fits perfectly because, just like long-distance running, writing a book is a test of endurance, discipline, and mental fortitude. You start with enthusiasm, hit several roadblocks along the way, feel the burn in the middle, but cross the finish line with an exhilarating sense of achievement.

When I was writing *Wealth Beyond Money*, the biggest challenge wasn't writer's block; it was finding the time. Between work, family, and countless other commitments, carving out time to write was tough. I had so many thoughts in my head, and I wanted to get them all out quickly. The problem was, I couldn't type fast enough, and organizing those thoughts was time-consuming.

And I didn't have AI to help me then. ChatGPT wasn't around–at least not to the general public–when I was writing *Wealth Beyond Money* in late 2021. ChatGPT was released November 30, 2022.

The Modern Writer's Room

Today, it's easier than ever to write a book quickly.

Did I use AI to help me write this book you're reading right now? Of course. I would be a hypocrite if I didn't.

But I didn't use AI to write the book *for* me. There is a big difference.

It is perfectly fine to use AI to help you. The concept of writers having help is nothing new. There's even a theory that Shakespeare was not one man, but a group of people.

The most famous writers have a team of people writing for and with them. Shonda Rhimes, Tom Clancy, and many others rely on writers' rooms and ghostwriters to help craft their stories.

You are the showrunner; let ChatGPT be your writers' room.

Using AI tools like ChatGPT is not cheating. It's leveraging modern technology to enhance your productivity and creativity. Just as writers' rooms are essential for TV shows and movies, AI can be an invaluable partner in your writing journey. It can help you brainstorm, organize, and refine your thoughts.

When I decided to write *ChatGPT to Double Your Business in 90 Days*, I knew I wanted to embody the principles I was teaching. One day, while running, an idea hit me, and I pulled out my phone to voice-dictate into a Google Doc. That raw, unfiltered thought became a key point in the book. Later, I used ChatGPT to polish it up, ensuring it captured my voice and message perfectly. It was like having a team of editors at my fingertips.

Why Write a Book?

Think of a book like a big business card. Being an author boosts your authority in whatever field you are in. It makes it easier to get more sales and higher-ticket sales. Humans trust people with authority.

A book helps organize your thoughts, gives you a reason to appear on shows, get media coverage, and makes it easier to secure speaking engagements.

Years ago, even though I was receiving great feedback as a speaker, I sometimes got ghosted when reaching out to new groups for engagements. Then one of my mentors said, "No one will take you seriously if you don't have a book." That was the spark I needed to finish *Wealth Beyond Money* and get it to market.

Wealth Beyond Money reached bestseller status in 50 different Amazon categories, including #1 bestseller in three categories, across multiple countries. This helped make things smoother for my speaking career, and the requests became inbound more than outbound. Suddenly, I was the one getting calls, not making them.

How to Write a Book Quickly, Ethically, and In Your Voice

First things first, mindset hacks:

- Your book doesn't have to be long. Attention spans are shorter than ever, so the shorter the better. Adopt the mindset of *The 90-Minute Book* by Dan Sullivan. Your goal is to provide value, not volume.
- Your book doesn't have to be perfect. Perfectionism can be paralyzing. It's better to have a good book published than a perfect book stuck in your head.

"Done is better than perfect." –Sheryl Sandberg

The Process

1. Start with an Outline or Table of Contents: Create a Google Doc and start with a simple outline. As thoughts come into your head, put them where they go in the outline.
2. Capture Your Thoughts: Use voice dictation on your phone to capture thoughts on the go. I've had some of my best ideas during a run, or in the shower. Don't worry about perfect phrasing; just get the idea down.
3. Organize Your Stories: In each section of the outline, write the points you want to make and think of a story or anecdote to support each point. My mentor Les Brown says, "Never make a point without a story, and never tell a story without a point."
4. Refine with AI: Use ChatGPT to help you edit and rewrite the rough drafts. Do this one chapter at a time. Make sure your custom instructions are on and that it is trained to speak in your voice.
5. Hire an Editor: Most publishers will help with editing, but if you are self-publishing, you can hire great affordable editors on Fiverr.com and Upwork.

Here's the exact ChatGPT prompt that I used to write this chapter:

> Act as an expert ghostwriter for NYT best-selling books.
> I want to write a chapter about [chapter topic]. Refer to my custom instructions and write in my voice. Expound on my rough ideas below. Use the point-story-point format. For each section, if I did not include a relevant real-world example or personal anecdote, please ask me for one.
>
> [paste your rough ideas here, including your personal examples and stories]

After reviewing the result, ask ChatGPT the follow-up question "What do you think may be missing from this chapter?" This could spark more thoughts and ideas that you might have forgotten about.

Remember, when using AI, always review the final output and check for accuracy and authenticity. When done right, your AI-assisted chapter should sound just like it would if you had done it the old way, agonizing over every word yourself.

For the other chapters in this book, I created a custom GPT, so that they all have a similar structure. I will teach you how to make custom GPTs later in this book.

Additional Tips ChatGPT Suggested I Include In This Chapter:

Detailed Outline Structure

Creating a detailed outline is the foundation of your book. Here's how to structure your outline:

- Introduction: Hook the reader with a compelling opening.
- Main Sections: Break down your content into clear, manageable sections.
- Subsections: Add detailed points and subpoints under each main section.
- Conclusion: Summarize key takeaways and leave the reader with a final thought.

Common book structures include the three-act structure for narratives or problem-solution frameworks for non-fiction.

Time Management and Writing Habits

Managing your time effectively is crucial:

- Set Dedicated Writing Time: Block out specific times each day for writing.
- Create a Routine: Develop a habit of writing at the same time each day.

Overcoming Writer's Block

Stay motivated and push through writer's block:

- Take Breaks: Sometimes stepping away helps.
- Change Your Environment: Write in different locations.
- Set Small Goals: Break your writing into manageable tasks.

Editing and Revisions

Editing is where the magic happens:

- Hire an Editor (see above)
- Multiple Drafts: Expect to write several drafts.
- Self-Editing Tips: Read your work aloud, take breaks between drafts, and focus on one type of edit at a time (e.g., grammar, structure).

Publishing Options

There are various ways to get your book published:

- Traditional Publishing: Submit your manuscript to publishers and potentially secure an advance.
- Self-Publishing: Maintain control and higher royalties but handle all aspects of publishing. I chose this option for *ChatGPT To Double Your Business In 90 Days*.
- Hybrid Publishing: A mix of both, offering professional services for a fee. I chose this option for *Wealth Beyond Money*.

Marketing Your Book

Marketing is essential to reach your audience. You should spend as much time (or probably more time) marketing your book as you did writing it.

- Build an Author Platform: Use social media, blogs, and podcasts to connect with readers.
- Leverage Email Lists: Collect emails and send updates about your book.
- Use Media: Reach out to media outlets for coverage.
- Get Help: Hire a book marketing professional.
- Document the Process: Start marketing the book while you are writing it.
- Make It an Event: Set a launch date, and get as many people as possible to buy it within 48 hours of launch. This is the trick to hitting bestseller lists.

Writing a book is a journey.

It might feel overwhelming at times, but with the right mindset and tools, it becomes an achievable goal. Remember, it's not about perfection; it's about progress. By leveraging AI, capturing your thoughts on the go, and organizing them effectively, you can produce a book that not only showcases your expertise but also opens doors to new opportunities. So, lace up your shoes, start running, and before you know it, you'll be crossing that finish line, book in hand, ready to inspire the world.

DAY 36: CREATING IMAGES WITH AI

AI image generation has tons of cool uses for businesses. It's perfect for spicing up marketing campaigns with unique visuals, creating eye-catching social media content, and designing custom product images. Real estate agents can use it for virtual staging, and e-commerce sites can show off products in different settings. Plus, it's great for whipping up website artwork, promo materials, and even book covers, saving time and money while keeping everything top-notch.

Creating images with AI involves using advanced machine learning models to generate visual content from text descriptions. This technology allows users to input detailed prompts and receive high-quality, custom images tailored to their specifications.

Researchers have been developing models that can understand and interpret language and visual data. These efforts culminated in the creation of models like OpenAI's DALL-E and other image generation tools. These models have been trained on vast datasets, learning to associate words with images to create accurate and creative visual representations from text prompts.

This technology saves time, enhances creativity, and allows for personalized visual content without needing advanced graphic design skills.

To create images using ChatGPT (DALL-E), follow these steps:

1. Ensure you have a subscription to the paid version of ChatGPT, which includes DALL-E.
2. Open ChatGPT and log in to your account.
3. Type a detailed description of the image you want to create in the chat. Start with "Create an image of..." For example: "Create an image of a serene beach at sunset with palm trees."
4. Submit the prompt and wait for ChatGPT to process and generate the image.
5. Check the generated image. If needed, refine your prompt for better results. The advantage of ChatGPT is that you can just continue the conversation to refine the image, instead of having to redo the entire prompt.

6. Right-click on the image and select "Save Image As..." to download it to your device.
7. Try different prompts and descriptions to create a variety of images.

To create images using Midjourney, follow these steps:

1. Visit Midjourney.com and follow the instructions.
2. Create a Discord account if you don't have one and join the Midjourney server. You may be able to create images directly in the browser in the future. At the time of this writing, this feature is only available to heavier users.
3. Subscribe to a Midjourney plan that fits your needs.
4. Find a "newbies" channel, which is designated for beginners.
5. In the text box, type /imagine prompt followed by a description of the image you want. For example: /imagine prompt: a futuristic city skyline at sunset.
6. The bot will generate four image variations based on your prompt. This usually takes about a minute.
7. Use the buttons under your image results to get upscale (U1, U2, U3, U4) or variation (V1, V2, V3, V4) options.
8. Choose which image you want to enhance or modify by clicking the respective button.
9. Once satisfied, click on the final image to view it in full resolution.
10. Right-click on the image and select "Save Image As..." to download it to your computer.
11. Experiment with different prompts and settings to get the best results.
12. Scroll through the community, the documentation, and other channels for inspiration and for advanced prompt ideas.

Some of the Top Tools for Creating Images With AI

- Midjourney: Renowned for its photorealistic results and user-friendly interface.
- DALL-E: Offers flexibility and creativity in generating unique images from detailed prompts.
- Stable Diffusion: Known for producing high-quality images and offering extensive customization options.

- Adobe Firefly: Integrates seamlessly with Adobe Creative Suite, making it a powerful tool for designers.
- Artbreeder: Allows for collaborative image generation and customization, ideal for creative projects.

Little Known Tips and Tricks

- Leverage Descriptive Prompts: The more detailed your prompt, the better the AI can understand and create your desired image. To reverse engineer this process, you can upload an image and ask AI to describe it for you.
- Use Style References: Mention specific art styles or famous artists to guide the AI in the direction you want.
- Experiment with Variations: Generate multiple versions of an image to explore different creative possibilities.
- Incorporate User Feedback: If using AI for client projects, gather feedback on initial images and refine prompts accordingly.

DAY 37: DESIGN YOUR BOOK COVER

When I first asked ChatGPT to "Create a book cover for ChatGPT To Double Your Business In 90 Days," it came up with this:

DALL·E 2024-01-22 10.44.00 - Book cover for 'ChatGPT to Double Your Business in 90 Days'. The cover features a sleek, modern design with a bold, eye-catching title in large, promi.png

As you can see, AI image generators aren't great at spelling. It is gradually getting better.

Using this mockup as inspiration, I added the text in Photoshop. Then I made a couple of other versions with the help of Midjourney.

I first used the same prompt in Midjourney, but it went in a completely different direction. Here is the original prompt and result:

/imagine prompt: book cover for "ChatGPT to double your business in 90 days" --ar 17:22 --v 6.0

These weren't even close to the look I wanted, so I changed it to this:

/imagine prompt: futuristic book cover art for "THE HANDS-ON AI AND AUTOMATION GUIDE TO EXPLOSIVE GROWTH IN TODAY'S ECONOMY" --ar 17:22 --v 6.0

Note: The --ar is for aspect ratio. 17:22 is the same as an 8.5" x 11" page. The --v 6.0 is telling Midjourney to use version 6, the most current version right now.

This was my favorite result from Midjourney:

I created a couple of different versions based on this.

I then did a poll on Facebook, Instagram, and LinkedIn to crowdsource opinions. I like this image, but several people commented that the robot hand looks "creepy."

To my surprise, Meta AI analyzed all of the comments and gave me this summary:

> **Meta AI**
>
> **What people are saying**
>
> The majority of commenters prefer cover B, citing that it looks more professional and eye-catching, while some prefer cover A for its brightness and clarity. A few suggest combining elements of different covers, like the title from C on cover B.

I used this feedback to help shape the final cover that you are holding.

I recommend this same approach with your book, because it makes your followers feel included, and gets them excited about what's to come.

DAY 38: CREATE LOGOS FROM SCRATCH WITH AI

There are a ton of AI logo generators, but my current favorite is Looka.

Just answer a few questions and it will generate dozens of beautiful logos in seconds, complete with a brand kit and mockups. It's free to look at the concepts, but if you want to buy the logo of course that costs money.

This can shortcut the process whenever you are launching a new product line or brand.

Try it free at EthanKing.com/Looka

DAY 39: USE YOUR SKETCHES TO CREATE LOGOS

Have an idea, but no graphic design experience?

1. Sketch it out.
2. Take a photo of it and upload it to ChatGPT (easiest with the ChatGPT mobile app).
3. In the same prompt box, type: "Make this a logo." See example below.
4. Continue the conversation to tweak the output. For example, you can say "Give me different options."

Make this a logo

Here's the new logo design for Simple Success Systems. If you have any adjustments or additional elements you'd like to see, let me know!

DAY 40: CREATE A WEBSITE IN SECONDS, SIMPLY BY DESCRIBING IT

Believe it or not, you can create an entire website from scratch, simply just typing one sentence.

Try it for yourself (free) at EthanKing.com/mixo

Your site will include a starter logo and pages with all of the text, stock images, and fake testimonials for you to replace later.

Just provide a brief description of your idea, and Mixo turns it into a complete website in seconds—no coding or design skills needed! It also offers landing pages, subscriber management tools, and customization options to connect with customers through email, surveys, and interviews. Perfect for pre-launching products and gathering insights.

Tools To Convert Leads Into Paying Clients

Ok, so now you've got a bunch of leads and that's great, but what's the next step? Converting those leads into paying clients is where the real magic happens. This chapter is packed with proven strategies and innovative tools designed to help you turn potential customers into loyal clients. Leveraging the latest in AI and automation, we'll dive into methods that streamline your sales process, making follow-ups and engagement more effective and efficient.

Drawing from personal experiences and industry insights, I'll guide you through practical applications of these tools. We'll explore automated follow-up sequences that ensure no lead is left behind, and AI-driven customer segmentation that personalizes your approach. These strategies not only enhance your conversion rates but also save you time and effort, allowing you to focus on what you do best. Get ready to transform those leads into committed clients and drive your business growth to new heights.

DAY 41: MASTER STORYTELLING SCIENCE

Storytelling is more than an age-old tradition; it's a powerful tool in business, especially when you weave it into your sales strategy.

Story*selling* is the strategic use of narrative to engage, persuade, and convert your audience. It's about turning your brand's message into a story that resonates deeply with your audience, compelling them to act.

The concept of storyselling has roots in the earliest forms of human communication. Storytelling has always been a way to convey information, values, and emotions. Over time, marketers and salespeople realized that the emotional connection forged through storytelling could be leveraged to influence buying decisions. In the digital age, where consumers are bombarded with information, a compelling story can cut through the noise and make a brand memorable.

Why It Works

Humans are wired to respond to stories. They trigger emotional responses and make information more memorable. According to psychologist Jerome Bruner, people are 20 times more likely to remember a fact when it has been wrapped in a story.[9] Stories engage the brain more than plain facts, tapping into our innate need for connection and meaning.

Leverage Vulnerability: Share Your Lows

Sharing personal lows and vulnerabilities can be surprisingly powerful. Vulnerability fosters trust and connection. When I shared some of my most shameful moments with a reporter, I never imagined the impact it would have. I spoke about how I used to take out the trash at a strip club and how a carjacking turned my life around. To my astonishment, this story led to me being featured on the front page of the AJC business section three times in one year. It was the most shared story in the "Secrets to Success" column.

---

[9] Vanessa Boris, "What Makes Storytelling So Effective For Learning?" *Harvard Business Publishing*, December 20, 2017, https://www.harvardbusiness.org/what-makes-storytelling-so-effective-for-learning/

This experience taught me that people connect with authenticity and vulnerability, and it's often the stuff we least want to share that resonates the most.

The Essential Elements of Compelling Content

Every piece of content should achieve at least one of the following:

1. Educate: Provide actionable "how-to" information.
2. Entertain: Engage with humor or captivating storytelling.
3. Inspire: Share transformation stories that motivate and uplift. Think "from-to."

By incorporating these elements, your content becomes something people want to share and save, increasing its reach and impact.

Effective Storytelling Frameworks

- Pain-Agitate-Solution: Identify your audience's pain points, agitate those pains by highlighting the consequences of inaction, and then present your product or service as the solution.
- Hero's Journey (Donald Miller, StoryBrand): Use this classic framework to map out a story where the customer is the hero and your brand is the guide that helps them overcome challenges.
- Epiphany Bridge Story (Russell Brunson, Expert Secrets): Share the moment of realization that led to your solution, helping your audience see the transformative power of your product.
- Money, Movement, Sex, Mastery (Tai Lopez): These four elements capture core human motivations and can be used to craft compelling narratives.

Some of the Top Tools for Storyselling

- ChatGPT: For generating and refining story ideas.
- StoryBrand's MyStoryBrand Tool: A free tool to help structure your hero's journey stories. (MyStoryBrand.com)
- Save The Cat: Screenwriting and storytelling resources (savethecat.com)
- BuzzSumo: To analyze and identify trending content topics (buzzsumo.com)

Little-Known Tips and Tricks

1. Let Others Tell Your Highs: It's more impactful when someone else shares your successes. It adds credibility and avoids coming across as boastful.
2. Show Movement from Low to High: Highlight the journey from struggle to success, focusing on the vehicle (your product or service) that facilitated this transformation.
3. Use Social Proof: Incorporate testimonials and case studies to reinforce your story's credibility.

Action Item

Let's revisit and update the blog post prompt from earlier to include these storytelling elements. Try this prompt in ChatGPT:

> Act as a compelling storyteller. Write a 500-word blog post on [topic] for [your website].
>
> It should be educational. Start the title with "How..." Do not include subheadings. Always communicate what benefits are in it for the reader.
>
> And it should have an inspirational tone that shows the **from-to** journey for the reader. Use the **pain-agitate-solution** framework where appropriate.
>
> Make it friendly, in my voice. Keep it grounded and on a 12th grade level. Include a story about how I went from [low point] to [high point] because of [product or service]. End with a call to action for the reader to (buy product, call you, leave a comment, etc.).

DAY 42: EMAIL DRIP CAMPAIGNS

Email drip campaigns are pre-scheduled sets of emails that are sent to contacts over a period of time based on specific triggers or tags. These sequences allow businesses to maintain consistent communication with leads, prospects, and customers without manual intervention. They can be tailored to various scenarios such as welcoming new subscribers, nurturing leads, following up on purchases, or re-engaging inactive customers.

The concept of automated email drip campaigns has its roots in early email marketing practices where businesses sought efficient ways to stay connected with their audience. With the advent of customer relationship management (CRM) systems and marketing automation software, the ability to schedule and personalize email sequences became more sophisticated.

Early pioneers like Infusionsoft (now Keap) and Mailchimp revolutionized how businesses approached email marketing by introducing tagging, segmentation, and automated workflows.

Why It Works

Automated email drip campaigns work because they ensure timely and relevant communication with your audience, fostering stronger relationships and increasing the likelihood of conversions. By automating routine follow-ups, businesses can focus on more strategic tasks while maintaining a consistent touchpoint with their customers. The personalization made possible by tagging and segmentation means that recipients receive content tailored to their specific interests and needs, enhancing engagement and driving action.

For instance at my company Zeus' Closet, consider a lead who has shown interest in custom artwork digitization. By tagging this contact and enrolling them in a "Digitizing" email sequence, we provide them with valuable information, examples of past work, and answers to common questions, all without manual follow-up. The lead receives an automated email from us every few days, but it doesn't look or feel automated. This personalized approach not only keeps the lead engaged but also positions us as attentive and responsive to their needs.

How to Do It Using AI to Help

AI can significantly enhance the effectiveness of your automated email drip campaigns by analyzing customer behavior, predicting preferences, and optimizing email content. Here's a step-by-step guide to setting up an AI-powered email sequence:

1. Identify Triggers and Tags: Use AI to analyze customer interactions and identify patterns that indicate interest or intent. For example, AI can track website visits, email opens, and clicks to determine the right moment to tag a contact for a specific sequence.
2. Personalize Content: AI can help craft personalized email content by analyzing customer data and generating tailored messages. Tools like ChatGPT can create compelling email copy that resonates with your audience. Try this prompt:

> Act as an expert digital marketer who specializes in getting high open rates.
>
> Write a 6-email campaign for [your company] to promote [specific product or service] to [target audience]. Here are some of the points we want to cover. Feel free to suggest others.
>
> The emails should be fun, personalized, educational, and witty, in our brand voice. We want to educate the reader as to why we are the best solution and how to work with us. Avoid words that might trigger spam filters.
>
> For each email, provide 3 options for curiosity-invoking subject lines in sentence case, an image suggestion, and a 200-word email body containing a story about one aspect of the topic and a call to action.

3. Optimize Timing: AI can predict the optimal times to send emails based on recipient behavior patterns, ensuring higher open rates and engagement.
4. Monitor and Adjust: Use AI to continuously monitor the performance of your email sequences, providing insights and recommendations for improvement.

Ways to Automate (or Semi-Automate) This Process

Automation tools like Zapier, Make.com and using Gmail labels can streamline the creation and management of email drip campaigns:

- Zapier or Make.com: Integrate Zapier with your email automation software to automate the tagging process. For instance, when a contact is added to a specific Gmail label, Zapier can automatically apply the corresponding tag in Keap, which triggers the appropriate email sequence.
- Gmail Labels and Tagging: Utilize Gmail labels to categorize contacts and streamline your tagging process. When a contact's label is updated, the change can trigger a Zapier automation that ensures they receive the right sequence.

Some of The Top Tools for Automated Email Drip Campaigns

- Klaviyo: Excellent for e-commerce businesses with robust automation and segmentation capabilities.
- ActiveCampaign: Known for its sophisticated automation workflows and predictive sending.
- Mailchimp: A user-friendly option with robust automation capabilities and extensive integrations.
- Drip: Ideal for e-commerce businesses, Drip provides targeted email marketing with deep customer insights.
- HubSpot: Offers comprehensive marketing automation tools with advanced segmentation and personalization features.
- Lemlist: Specializes in personalized email outreach, making it great for drip campaigns and cold emailing.
- SimpleSuccess.ai: Our soon-to-be-launched platform will provide powerful automation features tailored to your business needs, making it easier to set up and manage complex email sequences.

Little-Known Tips and Tricks

- Email Deliverability: Ensure your emails land in the inbox by regularly cleaning your email list, using double opt-in methods, and avoiding spammy language.

- Compelling Subject Lines: Craft subject lines that grab attention and entice recipients to open your emails. A/B testing different subject lines can help determine what works best. Avoid title case and overuse of emojis; they are a dead giveaway for AI
- Content Variability: Mix up your content with a combination of text, images, and videos to keep your audience engaged.
- Segmentation: Use advanced segmentation to create highly targeted email sequences that address the specific needs and interests of different customer groups. Our highly-segmented emails get more than 3x the open rates compared to the more general ones.

DAY 43: FOLLOWUP.CC

You can automatically send reminder emails to customers or prospects after a certain period, ensuring that no opportunity slips through the cracks. It's a method to maintain contact, remind customers of previous interactions, and encourage them to take the next step without requiring constant manual effort.

With businesses handling an increasing volume of leads and communications, manually tracking and following up with each contact eventually becomes impractical. Automation tools like Followup.cc were developed to address this challenge, allowing businesses to schedule and send reminder emails automatically based on predefined triggers and timelines.

In the hustle and bustle of daily operations, it's easy for both businesses and customers to lose track of emails and conversations. Reminder emails serve as gentle reminders that keep the dialogue going. This persistence demonstrates reliability and customer-centric service, which can significantly improve conversion rates and customer satisfaction.

In our business, we set up a series of automated reminder emails using Followup.cc to ensure we never miss out on potential sales. For instance, after sending a sales quote, we programmed Followup.cc to wait three days before sending a polite reminder email if we hadn't received a reply. This email reads:

Hey there,

We know you're busy...just wanted to make sure you received this. Let us know how we can help.

Thanks!

If there's still no response, another reminder goes out three days later:

Are you good with this, or want to make some changes?

And finally, after six more days, the last reminder is sent:

Still interested in this?

These automated emails have been incredibly effective. Customers often thank us for following up because they got busy or missed the initial emails. This persistence has helped us close numerous deals that might have otherwise fallen through the cracks.

How to Do It Using AI to Help

AI can enhance your auto follow-up strategy by personalizing and optimizing the content and timing of your emails. Here's how you can implement it:

1. Set Up Your System: Choose a tool like Followup.cc to manage your follow-ups. Configure it to trigger emails based on specific time intervals after the initial contact.
2. Draft Personalized Templates: Use AI tools like ChatGPT to help you draft personalized reminder email templates. AI can suggest wording that resonates with your audience and maintains a friendly, professional tone.
3. Analyze and Optimize: Use AI to analyze the performance of your reminder emails. Tools like HubSpot or Mailchimp can provide insights into open rates, response rates, and other metrics. AI algorithms can suggest the best times to send emails and recommend adjustments to improve engagement.

Some of the Top Tools for Auto Reminders

To stay ahead of the curve, consider using the following top-rated tools for auto reminders:

1. Followup.cc: Offers robust scheduling and automation features, perfect for streamlined follow-up sequences.
2. HubSpot Sales: Integrates with your CRM and provides powerful automation and analytics capabilities.
3. Mailchimp: Known for its user-friendly interface and advanced automation features.

4. Reply.io: Specializes in automated email outreach and follow-ups, with AI-driven personalization.
5. SalesLoft: Provides comprehensive sales engagement tools, including automated follow-up sequences.

Little Known Tips and Tricks

1. Test Different Timings: Experiment with different intervals between follow-ups to find what works best for your audience. For example, many people check email while at work, so 10AM-12PM or 1PM-3PM may be the best time to get email opens.
2. Use Multiple Channels: Combine email follow-ups with other channels like SMS or social media for a multi-touch approach.
3. Monitor Responses: Keep an eye on responses and adjust your follow-up sequences based on feedback and engagement levels.
4. Leverage AI Insights: Use AI tools to predict the best times to send follow-ups and optimize email content for higher engagement.

Action Item

To get started with your own auto reminder strategy, use this ChatGPT prompt to draft your first email:

> Draft a friendly and professional reminder email to a customer who received a sales quote three days ago but hasn't responded. Mention that you understand they might be busy and offer assistance with any questions or concerns they might have. Keep it short and fun.

Implementing a robust auto reminder system can transform your business by ensuring no lead is left unattended and every opportunity is maximized. With the help of AI and automation tools, you can maintain consistent communication, build stronger customer relationships, and ultimately drive more sales.

DAY 44: DIRECT MESSAGING

Direct messaging (DM) on social media refers to the private communication channel within social platforms where users can send messages directly to one another. Unlike public comments or posts, DMs offer a personal and direct way to engage with followers, potential customers, and clients. This one-on-one interaction can build stronger relationships, provide personalized customer service, and facilitate business transactions.

As social media platforms expanded, the need for private, instant communication grew. Platforms like Facebook, Instagram, Twitter, and LinkedIn integrated direct messaging features to enhance user experience and engagement. Initially, DMs were used mainly for casual interactions, but businesses quickly realized their potential for customer engagement and sales.

Direct messaging is effective for several reasons:

- Personal Connection: DMs create a sense of personal touch and exclusivity, making customers feel valued and heard.
- Immediate Response: It allows for real-time communication, which is crucial for addressing customer inquiries and closing sales quickly.
- Higher Engagement: People are more likely to open and respond to a DM than a public post or email.

In our business, we're constantly posting to our Instagram and Facebook stories. Whenever customers reply, their messages land in our DMs. From there, we typically ask if we can send them a quote via email and add them to our email list in exchange for discount codes.

How to Do It Using AI to Help

AI can significantly enhance your DM strategy by automating responses, personalizing messages, and analyzing data to improve engagement. Here's how:

- Automated Responses: Set up AI chatbots to handle common queries instantly, ensuring customers receive immediate replies even outside business hours.

- Personalization: Use AI to gather customer data and tailor messages based on their preferences, purchase history, and behavior.
- Sentiment Analysis: AI can analyze the tone of messages and alert you to urgent or sensitive issues that need a personal touch.

Example: Implementing AI-driven chatbots on your social media accounts can manage the initial interaction with customers who respond to your stories, automatically offering quotes or discount codes based on predefined criteria.

Ways to Automate (or Semi-Automate) This Process

- Auto-Response Tools: Platforms like ManyChat or Chatfuel can be integrated with Facebook and Instagram to automatically respond to comments and DMs.
- DM Campaigns: Use tools like MobileMonkey to create DM campaigns that automatically send messages to users who comment on your posts.
- Customer Relationship Management (CRM) Systems: Integrate social media with your CRM to keep track of interactions and automate follow-ups.
- SimpleSuccess.ai (coming soon) will auto reply to comments in your followers DMs.

Some of The Top Tools for Direct Messaging

- ManyChat: Excellent for creating automated DM responses and campaigns on Facebook and Instagram.
- MobileMonkey: Ideal for integrating with Facebook Messenger and Instagram DMs to automate conversations and lead generation.
- HubSpot: A comprehensive CRM that can integrate with social media platforms for managing DMs and customer interactions.
- Hootsuite: Manages all your social media in one place, including responding to DMs across different platforms.

Action Item

To get started with automating your DMs, try this prompt in ChatGPT:

> Create an automated response message for Instagram DMs that greets the customer, asks if they would like to receive a quote via email, and offers a discount code for joining our email list. Make it friendly and engaging.

Example output:

"Hi [Customer Name]! 👋 Thanks for reaching out! We'd love to provide you with a personalized quote. Can we send it to your email? Plus, join our email list today and enjoy a special discount on your next purchase! Just drop your email below to get started. 📧"

DAY 45: PROMOTIONAL PRODUCTS

The concept of promotional items dates back to the late 18th century when George Washington used commemorative buttons during his presidential campaign. However, the promotional products industry as we know it today began in the late 19th century when Jasper Meek, a newspaper printer, persuaded a local shoe store to distribute school bags with their logo printed on them. This idea quickly caught on, leading to the creation of a wide array of branded items used for marketing purposes.

Promotional items work because they create a tangible connection between the brand and the recipient. Unlike digital ads or email marketing, which can be fleeting, promotional items have a physical presence that keeps the brand top of mind for longer periods. When customers use a high-quality branded item, they associate the quality of the item with the quality of the brand. This creates positive brand reinforcement and increases the likelihood of repeat business and referrals.

At Zeus' Closet, we found that customers love our high-quality branded pens. These pens aren't just writing instruments; they're a daily reminder of our brand's commitment to excellence. Every time a customer uses our pen, they're reminded of the positive experiences they've had with our company.

Ways to Automate (or Semi-Automate) This Process

1. Automated Ordering: Set up automated ordering systems with suppliers to ensure a steady supply of popular promotional items. At Zeus' Closet, we can set up a branded online store for your business, so employees and fans can order merchandise branded with your logo. We handle all of the customer service and fulfillment for you.
2. CRM Integration: Integrate your Customer Relationship Management (CRM) system with your promotional item distribution. For instance, Salesforce can automate the process of sending out promotional items to new leads or loyal customers.

3. Fulfillment Services: Use fulfillment services that specialize in promotional items. Companies like Printful can handle the printing, storage, and shipping of your branded products automatically.

Some of the Top Tools for Promotional Items

- Promotique by Vistaprint: Offers a wide range of customizable promotional items.
- Zeus' Closet: Great for specialty custom apparel, and unique promotional products. Embroidered items can even be personalized one-offs. In full transparency, I am a co-owner of Zeus' Closet.
- Custom Ink: Known for their high-quality screen printed t-shirts and excellent customer service.
- 4imprint: A comprehensive platform for all types of promotional products.
- Zazzle: Ideal for personalized promotional items with an easy-to-use design tool.
- Alibaba: Great for bulk ordering and custom manufacturing.

Little Known Tips and Tricks

- Quality Over Quantity: Don't go cheap on promo products. Invest in high-quality items that customers will use regularly. A high-quality pen, like those from Zeus' Closet, can make a lasting impression.
- Subtle Branding: Unless you're a globally recognized brand, avoid large logos. Instead, focus on a design that resonates with the recipient, with your logo subtly included.
- Seasonal Items: Consider offering seasonal items that are timely and useful, like branded umbrellas in the rainy season or sunglasses in the summer.
- Unique Products: Stand out by offering unique promotional items that your competitors aren't using. Think outside the box with items like portable chargers, eco-friendly products, or fitness gear.

Action Items

- Identify Your Audience: Use tools like Google Analytics to understand your audience's demographics and preferences.
- Select the Right Items: Based on your audience analysis, choose promotional items that align with their interests and needs.
- Personalize Your Approach: Implement AI-driven personalization tools to customize promotional items for your recipients.
- Automate Distribution: Set up automated systems for ordering and distributing promotional items to streamline the process.

Try this ChatGPT prompt to brainstorm promotional items:

> ChatGPT, help me brainstorm high-quality promotional items that would appeal to my target audience of [insert your audience details], focusing on items they would use regularly and appreciate for their utility and design.

DAY 46: VIRTUAL AND IN-PERSON EVENTS

Virtual events took off with the rise of the internet and digital communication tools back in the late '90s and early 2000s. They really hit their stride in 2020 when businesses needed alternatives to physical gatherings. On the other hand, in-person events have been around for centuries, evolving from traditional fairs and community gatherings to the modern conferences and trade shows we see today.

Virtual and in-person events each have their unique strengths:

- Virtual Events: They're incredibly flexible, accessible, and cost-effective. Anyone can join from anywhere, and you can scale up to accommodate large audiences without worrying about physical space.
- In-Person Events: These offer a level of personal interaction and engagement that's hard to beat. There's nothing like the atmosphere of a live event, the chance to network in person, and the direct human connections that strengthen brand loyalty.

For instance, there was once a national sorority conference in Atlanta. Instead of blending in with other vendors at the convention center, we hosted a VIP event at our headquarters, complete with shrimp cocktails and champagne. Attendees left the conference to visit us. This exclusive setting created a memorable experience, and attendees were more inclined to shop and engage with our brand.

How to Do It Using AI

AI can supercharge both virtual and in-person events:

- Virtual Events: Use AI to personalize attendee experiences with chatbots, content recommendations, and real-time analytics. When I taught AI on Zoom, I leveraged AI tools to analyze engagement, provide instant feedback, and tailor content on the fly.
- In-Person Events: Streamline logistics and enhance attendee experience with AI. Think facial recognition for quick check-ins, sentiment analysis to read the room,

and AI-driven matchmaking for networking. At our VIP sorority event, AI could have tracked guest preferences, ensuring each attendee felt special.

Ways to Automate (or Semi-Automate) This Process

- Automated Invitations and Follow-Ups: Use AI-driven email marketing tools to send personalized invites and follow-ups based on attendee behavior and preferences.
- Event Management Platforms: Platforms like Eventbrite or Hopin integrate AI to manage registrations, ticketing, and tracking attendees seamlessly.
- Engagement Tools: AI chatbots and virtual assistants can handle queries, guide participants, and collect feedback during the event.
- Content Automation: AI tools can transcribe sessions, create summaries, and generate highlights for post-event distribution.

Some of The Top Tools for Virtual and In-Person Events

- Hopin: Perfect for virtual and hybrid events, with tools for networking, engagement, and analytics.
- Eventbrite: Great for managing both in-person and virtual events, with features for ticketing, registration, and promotion.
- Zoom: The go-to for video conferencing, webinars, and interactive virtual events.
- Bizzabo: An event management software that uses AI to enhance experiences through personalized recommendations and analytics.
- Brella: A networking tool that uses AI to match attendees with similar interests, optimizing networking opportunities.

Little Known Tips and Tricks

1. **Hybrid Event Strategies**: Blend virtual and in-person elements to cater to a wider audience. For example, live stream key in-person sessions for virtual attendees.
2. **AI-Driven Insights**: Use AI to analyze attendee data from past events to tailor future content and improve engagement.
3. **Engage with Gamification**: Add gamification elements to your events to boost participation and interaction.

4. **Leverage Social Media**: Use AI tools to manage and analyze social media engagement before, during, and after the event to maximize reach.

Action Item

Create a comprehensive event plan using ChatGPT. Try this prompt:

> Act as an expert event planner. Help me create a detailed plan for an event focused on [topic] for [target market]. Include strategies for both virtual and in-person elements, ways to enhance attendee engagement using AI, and suggestions for automating the event management process.

By integrating AI into your event planning and execution, you can create highly engaging and efficient experiences that leave a lasting impression on your attendees.

DAY 47: PERSONALIZED SMS CAMPAIGNS

You might ignore an email, but you'll usually at least read a text message...even if you don't act on it. Personalized SMS campaigns work wonders because they make each customer feel like a VIP.

In a sea of boring, generic marketing messages, a personalized text is like a breath of fresh air. Imagine getting a text that's just for you – it's impossible to ignore! Studies show that these tailor-made messages get way more attention. A study at Gartner found that SMS open rates can hit an impressive 98%, with most texts read within minutes.[10] This instant connection is why SMS is a game-changer for keeping your leads hooked and engaged.

How to Do It Using AI to Help

Leveraging AI, particularly ChatGPT, can significantly enhance the effectiveness of your SMS campaigns. Here's how:

1. Crafting Messages: Use ChatGPT to draft personalized messages that cater to different segments of your audience. For instance, if you run a fitness business, you can create various messages for new members, regular attendees, and those who haven't visited in a while.
   Example:
   - For new members: "Hi [Name], welcome to [Gym Name]! We're excited to have you. Don't forget your first class is on us. See you soon!"
   - For regular attendees: "Hey [Name], great job on sticking to your fitness goals! Here's a tip to enhance your workout this week: [Tip]. Keep it up!"
   - For inactive members: "Hi [Name], we miss you at [Gym Name]. Here's a special offer to get you back on track: [Offer]. Let's achieve those goals together!"

---

[10] Stanzie Cote, "The Future of Sales Follow-Ups: Text Messages," *Gartner,* October 4, 2019 https://www.gartner.com/en/digital-markets/insights/the-future-of-sales-follow-ups-text-messages

2. Scheduling and Targeting: Highlevel allows you to schedule these personalized messages and target specific groups based on their interactions, demographics, and behaviors. This ensures your messages reach the right people at the right time.

Ways to Automate (or Semi-Automate) This Process

Automation is key to scaling personalized SMS campaigns. Here's how you can automate or semi-automate the process:

1. Segment Your Audience: Use your CRM to segment your audience based on specific criteria such as purchase history, engagement level, and demographics.
2. Automate Message Creation: Integrate ChatGPT with your CRM to automatically generate personalized messages for different segments.
3. Schedule and Send: Use platforms like Highlevel to schedule these messages at optimal times for each segment.

Some of the Top Tools for Personalized SMS Campaigns

- Highlevel: A comprehensive marketing automation tool that helps schedule and send personalized SMS campaigns.
- Keap (formerly Infusionsoft): A CRM and marketing automation platform that supports SMS marketing.
- Twilio: A flexible communications platform that integrates with various CRM systems to send SMS messages.
- SimpleTexting: A user-friendly platform for managing SMS marketing campaigns with personalization features.
- EZ Texting: Another robust platform for SMS marketing that offers segmentation and personalization options.

Little-Known Tips and Tricks

- A/B Testing: Regularly test different messages to see which ones perform better. Use the insights to refine your campaigns.
- Personalization Tokens: Use tokens to insert personalized details such as the recipient's name, purchase history, or location.

- Engagement Triggers: Set up triggers based on user actions. For instance, if a lead clicks a link in your message but doesn't follow through, send a follow-up message.
- Shortcodes and Keywords: Use shortcodes and keywords to make it easy for leads to interact with your messages. For example, "Text JOIN to 12345 for exclusive offers."

Action Item

Here's a sample prompt to create a personalized SMS message using ChatGPT:

> Write a personalized SMS message for a customer who has not engaged with our fitness app in 30 days. Include an offer to entice them back and a motivational quote to inspire action.

DAY 48: RETARGETING ADS TURN BROWSERS INTO BUYERS

Retargeting, also known as remarketing, is a breakthrough in online advertising. It's that magic trick where your ad "follows" potential customers around the internet after they've visited your website. Imagine a friendly reminder popping up on their favorite sites or social media feeds, nudging them to come back and complete their purchase. It's like having a digital sales assistant who never sleeps.

Retargeting came to life in the early 2000s when marketers realized that most website visitors left without buying anything. Instead of waving goodbye, they wanted a way to bring those visitors back. By using cookies and tracking pixels, they could serve ads to people who had shown interest but hadn't yet made a purchase. This clever approach quickly proved its worth, turning casual browsers into paying customers.

Retargeting keeps your brand top-of-mind for potential customers. Let's face it, people rarely buy on their first visit to a website. They get distracted or want to compare options. Retargeting ads gently remind them about what they were interested in.

For instance, when we started retargeting at stuff4GREEKS, customers often told us, "I saw your ad on CNN.com!" This continuous exposure led to more sales and kept our brand visible. It made our company appear bigger than we really were, which established a sense of trust from our shoppers.

If you're not retargeting, you're probably leaving money on the table.

How to Do It Using AI to Help

1. Set Up Your Retargeting Campaign:
    - Choose a platform (Google Ads, Facebook Ads, etc.)
    - Install tracking pixels on your website to collect data on visitor behavior.
2. Segment Your Audience:
    - Use AI-powered tools to analyze visitor data and segment your audience based on behavior, interests, and demographics.

3. Create Compelling Ads:
   - Develop ads tailored to each segment. Use AI to generate copy and design elements that resonate with your target audience.
4. Monitor and Optimize:
   - Utilize AI for real-time analytics and A/B testing. Continuously refine your ads and targeting strategies based on performance data.

Ways to Automate (or Semi-Automate) This Process

Automation can streamline your retargeting efforts, saving time and ensuring consistent performance. Here are some ways to automate:

- Dynamic Retargeting Ads: Platforms like Google Ads and Facebook can automatically create ads featuring products a user viewed on your site.
- AI-Powered Audience Segmentation: Tools like Segment or Mixpanel can automatically group users based on behavior, making it easier to target them with relevant ads.
- Automated Ad Creation: Services like AdEspresso can automate the creation and testing of multiple ad variations to find the most effective combinations.

Some of The Top Tools for Retargeting

- Google Ads: Offers robust retargeting features, including dynamic retargeting and detailed audience insights.
- Facebook Ads: Known for its precise targeting options and extensive reach across Facebook and Instagram.
- AdRoll: Provides comprehensive retargeting solutions with cross-channel capabilities, including web, social media, and email.
- Criteo: Specializes in dynamic retargeting, offering personalized ads based on user behavior.
- SharpSpring Ads: An easy-to-use retargeting platform that integrates with major ad networks and provides detailed analytics.

Little Known Tips and Tricks

- Frequency Capping: Limit the number of times a user sees your ad to prevent ad fatigue.
- Sequential Messaging: Create a series of ads that tell a story or guide the user through the buying journey.
- Exclusion Lists: Exclude users who have already converted from seeing your retargeting ads, saving budget and improving ROI.
- Cross-Device Retargeting: Ensure your ads follow users across devices for a seamless retargeting experience.
- Customized Landing Pages: Direct retargeted users to personalized landing pages that match their previous interactions with your site.

Action Item

To get started with retargeting, use the following ChatGPT prompt to brainstorm ad ideas:

> Create a series of three retargeting ad copy variations for [Your Product] targeting users who visited the product page but did not complete the purchase. The ads should highlight key features, offer a discount, and address common objections.

By integrating retargeting ads into your marketing strategy, you can significantly boost your business's sales and conversion rates. Leverage AI and automation to optimize your campaigns and ensure you're capturing every potential customer who shows interest in your brand.

Remember, the key is persistence and continuous optimization—retargeting isn't a one-time effort but an ongoing process of engagement and refinement.

DAY 49: VOICEBOTS

Voicebot platforms like VoiceGenie are cutting-edge AI-powered systems that interact with users through voice commands and natural language processing. These platforms enable businesses to automate customer interactions, providing instant responses and delivering personalized experiences through human-like conversation.

They can handle a wide range of tasks, from answering frequently asked questions to processing orders and reservations, making them invaluable tools for boosting customer service and engagement.

Why It Works

Voicebot platforms are effective because they align with the growing consumer preference for quick, efficient, and hands-free interactions. Here's why they work:

- 24/7 Availability: Voicebots are always on, providing instant support to customers regardless of time zones or business hours.
- Consistency: They deliver consistent responses, ensuring that every customer receives the same high-quality service.
- Efficiency: By automating routine inquiries and tasks, voicebots free up human staff to focus on more complex and high-value activities.
- Personalization: Advanced voicebots can tailor their responses based on previous interactions, providing a personalized experience that builds customer loyalty.

How to Do It Using AI to Help

Implementing voicebot platforms like VoiceGenie in your business involves several steps, where AI can play a crucial role:

1. Identify Key Interactions: Determine which customer interactions can be automated. Use AI to analyze customer queries and identify common patterns and frequently asked questions.
2. Develop Conversational Flows: Create scripts and conversation flows that the voicebot will follow. AI can help optimize these flows by predicting customer intents and improving response accuracy.

3. Integration: Integrate the voicebot with your existing systems, such as CRM and order management systems. AI can assist in ensuring smooth data exchange and process automation.
4. Testing and Optimization: Use AI-driven analytics to monitor the performance of your voicebot. Continuously optimize its responses based on user feedback and interaction data.

Ways to Automate (or Semi-Automate) This Process

Automation can streamline the deployment and management of voicebot platforms. Here are a few strategies:

- Automated Setup Wizards: Platforms like VoiceGenie often come with setup wizards that guide you through the initial configuration.
- AI Training Modules: Use pre-trained AI models that understand industry-specific language and terms, reducing the time required for custom training.
- Scheduled Updates: Automate updates and improvements to your voicebot's conversational capabilities based on periodic analysis of interaction data.
- Self-Learning Mechanisms: Implement AI that enables the voicebot to learn from each interaction, gradually improving its performance without manual intervention.

Some of The Top Voicebot Platforms

- VoiceGenie: Known for its robust features and ease of integration.
- Google Dialogflow: Offers powerful NLP capabilities and integrates well with various platforms.
- Amazon Lex: Backed by AWS, it provides scalable voicebot solutions.
- Microsoft Azure Bot Service: Combines with Azure's AI and machine learning tools for enhanced performance.
- IBM watsonx Assistant: Renowned for its AI-driven conversational capabilities.

Little-Known Tips and Tricks

- Utilize Multi-Channel Support: Ensure your voicebot can handle interactions across multiple channels, including phone, web, and mobile apps.

- Personalization Tokens: Use personalization tokens to insert customer-specific details into responses, enhancing the user experience.
- Regular Audits: Conduct regular audits of your voicebot's performance to identify and address any shortcomings.
- Voice Tone and Style: Customize the tone and style of your voicebot to match your brand's personality, creating a more cohesive customer experience.

Action Item

To help you get started with voicebot platforms, try using the following ChatGPT prompt:

> Create a conversational flow for a voicebot that handles customer inquiries for an e-commerce business. The flow should include greetings, product information requests, order status checks, and handling of common issues such as returns and refunds. Ensure the voicebot can escalate complex queries to a human representative.

DAY 50: AUTOMATED SURVEY FOLLOW-UPS

Automated survey follow-ups are key for businesses. Imagine this: after you've made initial contact with a potential lead, you send them a quick survey to gather their feedback. The best part? It's all automated.

The responses help you tailor your next steps, making your follow-ups super relevant and personal. It's like having a conversation with each lead, but without the manual effort.

Surveys have been around forever, right? Businesses have always wanted to know what their customers think. But automating this process? That's where the magic happens. Thanks to advancements in technology and AI, we can now streamline the entire process, ensuring timely and relevant communication with our leads. This modern twist on an old concept has transformed how we engage with potential customers.

So, why does this work so well? It's simple: people love to feel heard.

By asking for feedback, you show your leads that you care about their opinions. This builds trust and rapport.

Plus, the insights you gain from these surveys allow you to send follow-up messages that are spot-on. When leads see that you understand their needs, they're more likely to convert. It's all about making them feel valued and understood.

How to Do It Using AI to Help

1. Set Up Your Survey: Start by creating a survey with questions that matter to your business and your leads. Some CRMs like Highlevel have this built-in. If not, you can use something like JotForm, which integrates smoothly with most CRMs, to keep everything in one place.
2. Automate the Sending Process: Schedule your survey to go out 24-48 hours after your initial contact. This timing is key – it keeps you fresh in their mind without being too pushy.

3. Analyze the Responses: Let AI do the heavy lifting. Use AI to analyze the responses and uncover patterns you might miss.
4. Personalize Follow-Up Messages: Use the insights from the survey to craft personalized follow-up messages. Highlevel's AI capabilities can help suggest content that really hits home with each lead.

Ways to Automate (or Semi-Automate) This Process

- Integration with CRM: Ensure your survey software integrates with your CRM so that survey responses automatically update lead profiles.
- Email Automation: Use marketing automation software to automate your follow-up emails based on survey responses.
- AI Analysis: Employ AI tools for deep dives into your survey data, giving you actionable insights.
- Template Creation: Develop email templates that can be easily personalized based on survey feedback, ensuring consistency and saving time.

Some of the Top Tools for Automated Survey Follow-Ups

- JotForm: Excellent for creating customized surveys and integrates well with various CRMs.
- Highlevel: Robust platform for automating surveys, follow-ups, and analyzing responses with AI capabilities.
- Typeform: Engaging, conversational surveys that feel less like a chore.
- Qualtrics: Advanced analytics and excellent CRM integration.
- SurveyMonkey: User-friendly and integrates well with various platforms.

Little Known Tips and Tricks

- Use Conditional Logic in JotForm: Make your surveys smarter by adjusting questions based on previous answers.
- Incentivize Participation: Offer small rewards like discounts or freebies to boost survey completion rates.
- Keep it Short: Respect your leads' time with brief, to-the-point surveys.

- Test and Optimize: Regularly experiment with different questions and follow-up messages to find what works best.

Action Item

Try this prompt:

> ChatGPT, help me create a follow-up email based on the following survey responses: [insert survey responses]. Make the email engaging and tailored to the lead's specific needs.

By leveraging automated survey follow-ups, you're not just streamlining your process – you're creating a more personalized and engaging experience for your leads. It's all about making meaningful connections and driving business growth. Give it a try and watch your conversion rates soar!

DAY 51: COPYWRITING FOR WEBSITE PAGES

Crafting effective copy for your website involves writing content that appeals to your readers while also being optimized for search engines (SEO). This means creating compelling, engaging, and informative text that encourages visitors to stay on your site, explore your products or services, and ultimately take action, such as making a purchase or signing up for a newsletter. At the same time, the content should include relevant keywords and phrases to improve your site's ranking on search engine results pages (SERPs).

Key Elements of Website Copywriting

- Audience-Centric: Understand your audience and write content that speaks directly to their needs, desires, and pain points.
- SEO-Optimized: Incorporate relevant keywords naturally to improve search engine visibility.
- Clear and Concise: Use straightforward language and avoid jargon to ensure your message is easily understood.
- Compelling CTAs: Include strong calls-to-action (CTAs) to guide users towards desired actions.
- Value Proposition: Clearly communicate the benefits and unique selling points of your products or services.
- Consistency: Maintain a consistent tone and style that aligns with your brand's voice.

ChatGPT Prompts for Help with Copywriting

ChatGPT, I need assistance with copywriting for my website pages, including landing pages and sales letter pages. Please help me craft engaging, SEO-optimized content that appeals to my target audience and includes strong calls-to-action. My website focuses on [your business niche/industry]. Here's an overview of what I need:

1. Homepage: A brief introduction to my business, highlighting our unique value proposition.
2. About Us Page: A compelling narrative about our brand, mission, and values.
3. Product/Service Pages: Detailed descriptions of our products/services, emphasizing benefits and features.
4. Landing Pages: Persuasive content for our marketing campaigns, designed to convert visitors into leads or customers.
5. Sales Letter Pages: A direct and persuasive pitch to encourage purchases or sign-ups.

Include relevant keywords such as [list your main keywords] to improve SEO. Let's start with the homepage content.

Act as an expert copywriter specialized in writing website copy.

Optimize the copy on this web page, making sure that it is highly engaging and readable, contains the most relevant keywords for SEO, and conveys my business' unique value propositions so that it turns visitors into customers.

[paste the URL here]

DAY 52: PERSONALIZED AI CHATBOTS

When it comes to transforming your business, few tools have the potential of personalized AI chatbots. Imagine a system that not only interacts with your customers but also understands their needs, preferences, and behaviors.

This isn't just customer service; it's personalized customer engagement, guiding potential clients through your sales funnel with the precision of a seasoned salesperson. In this chapter, we'll explore the concept of personalized AI chatbots, their origins, why they work, and how you can leverage them using AI.

Unlike generic chatbots that offer standard responses, personalized AI chatbots learn from each interaction, using data to tailor responses and recommendations to individual users. These chatbots can recommend products, answer specific questions, and even remember past interactions, creating a seamless and engaging customer experience.

The idea of chatbots isn't new. They date back to the 1960s with ELIZA, one of the first chatbots developed by MIT's Joseph Weizenbaum. However, the concept of personalization in chatbots gained traction with advancements in machine learning and natural language processing (NLP). As AI technology evolved, so did the capability of chatbots to understand and respond to human language in more nuanced ways, leading to the sophisticated, personalized chatbots we use today.

Why It Works

Personalized AI chatbots work because they cater to the individual needs of each customer. In a world where consumers expect immediate and relevant responses, a chatbot that can provide personalized interactions stands out. These chatbots enhance customer satisfaction by reducing wait times and providing accurate information, which in turn boosts engagement and conversion rates.

Ways to Automate (or Semi-Automate) This Process

Automation is key to maximizing the efficiency of personalized AI chatbots. Here are some ways to achieve this:

1. Automated Responses: Set up your chatbot to handle common queries automatically. This frees up human agents to tackle more complex issues.
2. Lead Qualification: Use chatbots to qualify leads by asking pertinent questions and scoring their responses based on predefined criteria.
3. Follow-Up Reminders: Automate follow-up messages to customers who showed interest but didn't complete a purchase, nudging them gently down the sales funnel.

Some of The Top Tools for Personalized AI Chatbots

Selecting the right app for building your chatbot depends on several factors including ease of use, integration capabilities, scalability, and cost. Here are some of the top chatbot platforms you can consider, along with their features to help you make an informed decision:

ManyChat

Features:

- Ease of Use: User-friendly drag-and-drop interface.
- Integrations: Seamlessly integrates with Facebook Messenger, Instagram, and other social media platforms.
- Templates: Offers pre-built templates for various industries.
- Automation: Supports automation and broadcasting.
- Follow-Up: Built-in tools for follow-up messages and email collection.

Best For:

- Businesses looking for a straightforward and social media-focused chatbot solution.

Dialogflow (by Google)

Features:

- AI-Powered: Leverages Google's natural language understanding.
- Multichannel: Supports integration with multiple platforms including web, mobile, and social media.
- Customization: Highly customizable with advanced capabilities.
- Follow-Up: Supports complex follow-up questions and flows.

Best For:

- Businesses needing a highly customizable and powerful AI chatbot.

Tars

Features:

- Ease of Use: Drag-and-drop builder with no coding required.
- Templates: Offers numerous templates for different use cases.
- Integrations: Integrates with CRM systems, Google Analytics, and more.
- Follow-Up: Built-in follow-up and re-engagement features.

Best For:

- Businesses looking for a simple yet effective solution for lead generation and customer service.

Chatfuel

Features:

- Ease of Use: User-friendly interface with drag-and-drop functionality.
- Templates: Pre-built templates available for various industries.
- Integrations: Supports integration with Facebook Messenger, Instagram, and other platforms.
- Follow-Up: Allows for automated follow-up messages and lead nurturing.

Best For:

- Businesses focusing on Facebook and Instagram marketing.

Botsify

Features:

- Ease of Use: Intuitive interface with drag-and-drop builder.
- AI-Powered: Supports machine learning for improved responses.
- Integrations: Integrates with websites, social media, and messaging apps.
- Follow-Up: Allows for scheduled follow-up messages and email collection.

Best For:

- Businesses looking for an affordable, easy-to-use chatbot with good support.

ManyChat or Chatfuel would be highly suitable if you're focusing on social media interactions, particularly on Facebook and Instagram.

If you require a more customizable and AI-driven solution with broader integration capabilities, Dialogflow would be the best choice.

Implementation Steps

1. Sign Up: Create an account on the chosen platform.
2. Define Objectives: Clearly outline what you want the chatbot to achieve.
3. Build the Chat Flow: Use the drag-and-drop interface to create your chat flow based on the guide provided earlier.
4. Integrate: Connect the chatbot with your website or social media platforms.
5. Test: Run multiple scenarios to ensure it works smoothly.
6. Launch: Deploy the chatbot and start engaging with your visitors.
7. Optimize: Collect feedback and continuously improve the chatbot experience.

DAY 53: CREATE PROPOSALS

Why reinvent the wheel? Get ChatGPT to draft your initial proposal, then you can refine it.

Drafting proposals can be time-consuming, but leveraging AI like ChatGPT can streamline the process. Instead of starting from scratch, you can use ChatGPT to generate a strong initial draft, which you can then refine to meet your specific needs. This method saves time, ensures a professional tone, and provides a solid foundation that you can build upon.

ChatGPT Prompt for Creating Proposals

> Hi ChatGPT,
> I need help drafting a proposal for [specific purpose or project]. Here are the details:
> - Objective: [Describe the main goal of the proposal]
> - Background: [Provide some context or background information]
> - Scope: [Outline what will be covered in the proposal]
> - Key Points:
>   - [First key point]
>   - [Second key point]
>   - [Third key point]
> - Benefits: [Explain the benefits of the proposal]
> - Timeline: [Include a proposed timeline for the project]
> - Budget: [Provide a rough estimate of the budget]
>
> Please structure the proposal professionally and include an introduction, main body, and conclusion.
> Thank you!

> I just had a call with a potential client who is having [problem] and wants to [outcome]. Here are the services I am offering her: [list services]
> Research other businesses in my industry and draft a proposal that includes scope, deliverables, price, timeline, terms, and any other relevant information.

DAY 54: AUTOMATED SALES WORKFLOWS

An automated sales workflow is a system that uses technology to manage and streamline the various stages of the sales process. This includes tasks such as lead generation, follow-ups, proposal generation, and contract management. By automating these tasks, businesses can save time, reduce human error, and ensure a consistent and efficient sales process.

The concept of automating sales workflows emerged from the broader trend of business process automation (BPA) in the late 20th century. As companies began to adopt CRM systems and digital tools, the potential to automate repetitive sales tasks became apparent. The advent of AI and machine learning in the 2010s further revolutionized this field, allowing for more sophisticated and intelligent automation solutions.

Why It Works

Automating sales workflows works because it addresses several common pain points in the sales process:

- Consistency: Automated workflows ensure that every lead and customer receives the same high level of attention and follow-up, reducing the risk of missed opportunities.
- Efficiency: Automating repetitive tasks frees up sales teams to focus on higher-value activities, such as building relationships and closing deals.
- Accuracy: Automation reduces the risk of human error in tasks such as data entry, proposal generation, and contract management.
- Scalability: As businesses grow, automated workflows can scale with them, handling an increasing volume of leads and customers without a corresponding increase in manual work.

How to Use AI to Help

AI can enhance sales workflows in several ways:

1. Lead Scoring and Prioritization: AI algorithms can analyze vast amounts of data to score leads based on their likelihood to convert. This helps sales teams prioritize their efforts on the most promising prospects.
2. Personalized Communication: AI can analyze customer data to generate personalized follow-up emails and messages, increasing engagement and conversion rates.
3. Proposal Generation: AI tools can create tailored proposals based on customer data and preferences, saving time and ensuring accuracy.
4. Contract Management: AI-powered contract management systems can automate the creation, review, and approval of contracts, reducing delays and errors.

Ways to Automate (or Semi-Automate) This Process

- Email Marketing Automation: Tools like Mailchimp or HubSpot can automate email follow-ups, nurturing leads through personalized drip campaigns.
- CRM Systems: Platforms like Salesforce or Zoho CRM can automate many aspects of the sales process, from lead management to deal tracking and reporting.
- AI Chatbots: Implementing AI chatbots on your website can help capture and qualify leads 24/7, providing instant responses to common queries.
- Proposal Software: Tools like Proposify or PandaDoc can automate the creation and tracking of sales proposals, integrating with your CRM to pull in relevant data.

Some of the Top Tools for Automated Sales Workflows

- HubSpot Sales Hub: Known for its comprehensive suite of tools, HubSpot offers everything from email tracking to meeting scheduling and pipeline management.
- Salesforce Einstein: Salesforce's AI-powered platform provides advanced analytics, lead scoring, and personalized recommendations.
- PandaDoc: This tool simplifies proposal and contract creation with templates and integrations with various CRMs.

- Outreach: A sales engagement platform that automates and personalizes outreach across multiple channels.
- Drift: A conversational AI platform that uses chatbots to engage website visitors and capture leads in real-time.

Little-Known Tips and Tricks

- Leverage AI Analytics: Use AI analytics to continually refine and optimize your automated workflows based on performance data.
- Integrate Tools: Ensure all your sales tools are integrated for seamless data flow and a unified view of your sales pipeline.
- Customize Automation: Tailor your automation workflows to your specific sales process and customer journey for maximum effectiveness.
- Regular Updates: Keep your automation tools and workflows up to date with the latest features and best practices.

Action Item

Use this ChatGPT prompt to help you build out your automated sales workflow. Make sure your custom instructions are populated with your business information, and are turned on:

> Act as an experienced sales manager at a high-performing firm. Given what you know about my business, what do you think is the sales workflow? And what are some ways we can automate much of it?

DAY 55: ABANDONED CART EMAILS

Abandoned cart emails are follow-up messages sent to potential customers who have added items to their online shopping cart but left the website without completing the purchase. These emails aim to remind the customer of their unfinished transaction, encouraging them to return to the site and complete their purchase.

The concept of abandoned cart emails emerged with the rise of e-commerce. As online shopping became more prevalent, retailers noticed a significant number of customers who would add items to their carts but leave without completing the transaction. Recognizing a lost sales opportunity, marketers devised a strategy to recover these sales through automated follow-up emails. The success of these emails in recapturing potential revenue solidified their place in digital marketing strategies.

Why It Works

Abandoned cart emails work because they address a crucial moment in the customer journey. When a shopper adds items to their cart, they have already expressed a strong interest in the product. However, various factors such as distractions, doubts, or technical issues can prevent them from completing the purchase. An abandoned cart email serves as a gentle nudge, reminding them of their intended purchase and often providing incentives like discounts or free shipping to sweeten the deal.

A real-world example of this is our own experience at stuff4GREEKS. By implementing an abandoned cart email strategy, we were able to recover 19.05% of our abandoned orders this year. This significant recovery rate translates directly into increased revenue and demonstrates the effectiveness of this strategy.

How to Use AI to Help

Artificial Intelligence can significantly enhance the effectiveness of abandoned cart emails. AI-powered tools can analyze customer behavior and preferences to personalize the content of the emails. For instance, AI can recommend related products, offer personalized discounts, or adjust the timing of the email based on when the customer is most likely to open it.

In addition, AI can segment customers based on their behavior and engagement, allowing for more targeted and effective follow-up strategies. By leveraging machine learning algorithms, businesses can predict which customers are more likely to respond to abandoned cart emails and tailor their approaches accordingly.

Some of The Top Tools for Abandoned Cart Emails

- Klaviyo: Known for its robust automation features and powerful segmentation capabilities, Klaviyo is a top choice for abandoned cart emails. It integrates seamlessly with e-commerce platforms like Shopify and WooCommerce.
- Mailchimp: With its user-friendly interface and comprehensive automation options, Mailchimp remains a popular choice for businesses of all sizes. It offers customizable abandoned cart email templates and detailed analytics.
- Omnisend: Omnisend specializes in e-commerce email marketing and offers advanced automation workflows, including abandoned cart emails. It also provides SMS marketing options for multi-channel engagement.
- HubSpot: HubSpot's marketing automation platform includes abandoned cart email functionality along with powerful CRM integration, allowing for highly personalized follow-ups.
- Sendinblue: This platform combines email marketing with SMS marketing, making it a versatile tool for abandoned cart recovery. Its automation features are designed to maximize engagement and conversion rates.

Little Known Tips and Tricks

- Send Multiple Reminders: Rather than sending just one abandoned cart email, consider a series of reminders. The first email can be sent within an hour of cart abandonment, followed by additional emails over the next few days.
- Include Customer Reviews: Adding positive customer reviews or testimonials in your abandoned cart emails can build trust and encourage hesitant shoppers to complete their purchase.
- Test Subject Lines: A compelling subject line can significantly impact open rates. Experiment with different subject lines to see what resonates best with your audience.
- Use Urgency and Scarcity: Create a sense of urgency by mentioning limited stock availability or time-sensitive discounts. This can prompt customers to act quickly.
- Offer Assistance: Sometimes, customers abandon their carts due to questions or concerns. Offering a direct line to customer support or a chatbot in your email can help address these issues and facilitate the purchase.

Action Item

Try this ChatGPT prompt:

Act as an e-commerce expert. Draft a series of three abandoned cart emails for [your business]. Write in my brand voice. The first email should be a friendly reminder, the second should offer a small discount, and the third should create a sense of urgency with a time-sensitive offer. Include subject lines for each email and ensure the tone is engaging and supportive.

(wait for response)

Great. What are we missing?

DAY 56: MOTIVATING YOUR SALES TEAM

Motivating your sales team is crucial for maintaining high levels of productivity, morale, and overall success. A motivated sales team is more likely to achieve targets, show better job satisfaction, and contribute positively to the company's growth. Key strategies to motivate your sales team include setting clear and achievable goals, recognizing and rewarding performance, providing opportunities for growth and development, fostering a positive and supportive work environment, and offering competitive incentives.

ChatGPT Prompts

> Help me develop a comprehensive strategy to motivate my sales team. I want to include methods for setting clear goals, recognizing and rewarding performance, providing growth opportunities, fostering a positive work environment, and offering competitive incentives. Please provide actionable steps and examples for each method.

> Act as an expert Sales Manager.
> Here is how I currently manage my sales team each week: [briefly describe how you track performance, praise, reward, etc.]. What can I do to motivate my sales team to do better?

> When my sales team feels defeated, what's the best thing I can do to help them overcome this?

> My sales team's next goal is [goal].
> What's the best way to keep them motivated to reach that goal?

DAY 57: COMPETITIVE ANALYSIS

Competitive analysis involves a thorough examination of your competitors to understand their strengths, weaknesses, brand positioning, and sales performance. This analysis helps you identify the strategies they are using and gain insights that can inform your own business strategies.

By understanding where your competitors excel and where they falter, you can better position your brand in the market, capitalize on their weaknesses, and adapt successful tactics to improve your own performance.

Key Components of Competitive Analysis:

1. Strengths: Identify what your competitors do well, such as product quality, customer service, or marketing strategies.
2. Weaknesses: Determine areas where your competitors are lacking, which could be opportunities for your business.
3. Brand Positioning: Understand how your competitors are perceived in the market and what unique value propositions they offer.
4. Sales Performance: Analyze their sales figures, growth trends, and market share to gauge their financial health and success.

ChatGPT Prompts for Competitive Analysis:

> Give me a step-by-step guide on how I can find the best marketing strategies my competitors in the [niche] are using with their customers.

I am conducting a competitive analysis for my business. Could you help me gather information on the following competitors:

1. Competitor 1:
    - Strengths
    - Weaknesses
    - Brand Positioning
    - Sales Performance
    - Strategies they are using
2. Competitor 2:
    - Strengths
    - Weaknesses
    - Brand Positioning
    - Sales Performance
    - Strategies they are using
3. Competitor 3:
    - Strengths
    - Weaknesses
    - Brand Positioning
    - Sales Performance
    - Strategies they are using

Please provide detailed insights for each aspect and suggest any observable strategies that I could consider adopting or countering.

DAY 58: NEW PRODUCT RESEARCH & DEVELOPMENT

New Product Research & Development (R&D) involves the process of discovering new product opportunities, refining product concepts, and developing them into market-ready products.

This process typically includes market research, competitor analysis, ideation, prototyping, testing, and refining based on feedback. The goal is to create innovative products that meet consumer needs and stand out in the market.

ChatGPT Prompt:

> Act as my research & development team. Help me brainstorm new products for [your business]. Search our site for what we already offer, then suggest 3 new products that you think will perform well with our target market, based on actual research. Provide reasons to support your suggestions.

Let's say that out of ChatGPT's responses, you like idea #2, but don't know where to start. Follow up with this prompt next:

> Great! Teach me step-by-step how to do #2

DAY 59: MAKE A PITCH DECK TO RAISE CAPITAL

A pitch deck is a brief presentation used to provide your audience with a quick overview of your business plan. It's typically used during face-to-face or online meetings with potential investors, customers, partners, and co-founders.

A pitch deck usually consists of 10-20 slides and showcases your company's vision, product, business model, team, financials, and growth potential. The goal is to generate interest and secure a follow-up meeting or investment. It helps to have a massive, impactful goal that you can't achieve on your own.

Essential Components of a Pitch Deck

1. Introduction: Briefly introduce your company and what it does.
2. Problem: Explain the problem your business solves.
3. Solution: Describe your product or service and how it addresses the problem.
4. Market Opportunity: Show the size and potential of your target market.
5. Product: Highlight key features and benefits of your product or service.
6. Business Model: Outline how your business makes money.
7. Go-to-Market Strategy: Explain your plan for reaching customers and achieving growth.
8. Traction: Provide evidence of your progress and success to date.
9. Competition: Identify your competitors and explain your competitive advantage.
10. Team: Introduce your team and their relevant experience.
11. Financials: Present your financial projections and key metrics.
12. Funding Ask: Specify the amount of funding you're seeking and how you plan to use it.
13. Closing: Summarize the key points and end with a strong closing statement.

ChatGPT Prompt to Help Create a Pitch Deck

> Help me create a pitch deck to raise capital for my business. I want to raise money from investors for [describe your massive goal here]. I need guidance on what to include in each slide. Here are the details about my company:
> - Company Name: [Your Company Name]
> - Industry: [Your Industry]
> - Problem: [Describe the problem your business solves]
> - Solution: [Describe your product or service]
> - Market Opportunity: [Provide details on your target market size and potential]
> - Product Features: [Highlight key features and benefits]
> - Business Model: [Explain how your business makes money]
> - Go-to-Market Strategy: [Outline your plan for reaching customers]
> - Traction: [Provide evidence of progress and success]
> - Competition: [Identify competitors and your competitive advantage]
> - Team: [Introduce your team and their experience]
> - Financials: [Present financial projections and key metrics]
> - Funding Ask: [Specify the amount of funding needed and its intended use]
>
> Please organize this information into a compelling pitch deck with clear, concise, and engaging slides.

Follow up with this:

> Great! Now actually make the slides.

Believe it or not, ChatGPT will actually create a PowerPoint presentation or PDF for you to download!

Tools to Increase Average Order Value

Alright, friends, let's talk about leveling up your business game by boosting your average order value (AOV). You've done the hard work of getting customers to your store, so why not maximize each transaction?

Think about it: every time a customer hits that checkout button, there's an opportunity to add more value—not just for them, but for your business too. Whether it's through strategic upselling, bundling products, or offering premium service tiers, the tools we're diving into in this chapter will show you how to transform a simple purchase into a full-fledged experience that leaves your customers thrilled and your revenue soaring.

In this chapter, we're going to explore the art and science of increasing your AOV using AI and automation. From dynamic pricing models to irresistible product bundling and personalized upsells, these strategies aren't just about making more money—they're about delivering more value and enhancing the customer journey. By the end of this chapter, you'll have a toolkit of techniques that not only boost your bottom line but also foster stronger customer relationships and loyalty. So buckle up, because we're about to turn every sale into a win-win for both you and your clients!

DAY 60: DYNAMIC PRICING MODELS

Dynamic pricing, also known as demand pricing or time-based pricing, is a strategy where prices are adjusted in real-time based on various factors such as demand, competition, and customer behavior. This approach is commonly used in industries like airlines, hospitality, e-commerce, and ride-sharing services.

For instance, flight prices fluctuate based on factors like seat availability, booking timing, and competitor pricing. Similarly, platforms like Amazon adjust product prices dynamically to stay competitive and optimize sales.

By leveraging AI, businesses can implement sophisticated dynamic pricing models that analyze vast amounts of data to make real-time pricing decisions. This allows companies to maximize revenue, improve market competitiveness, and better respond to changing market conditions.

ChatGPT Prompt to Help with Dynamic Pricing Models

> Hi ChatGPT, I need assistance in implementing dynamic pricing models for my business. Could you help me understand the key steps and considerations for setting up a dynamic pricing strategy? Specifically, I need to know:
> 1. How to gather and analyze data on market demand, competitor prices, and customer behavior.
> 2. The best AI tools or platforms for implementing dynamic pricing.
> 3. Examples of successful dynamic pricing implementations in various industries.
> 4. How to monitor and adjust pricing strategies over time to ensure optimal performance.
> 5. Potential challenges and how to address them.
>
> Your guidance on these aspects will help me optimize my product pricing based on real-time market conditions. Thanks!

DAY 61: PREMIUM SERVICE TIERS

Offering premium service tiers is a strategy to provide your most valued customers with exclusive benefits and content, thereby enhancing their experience and increasing their loyalty. This approach involves creating different levels of service, with each tier offering unique perks. The highest tiers are typically reserved for high-spending customers and come with the most exclusive benefits.

By leveraging AI, you can efficiently manage these premium tiers.

AI can help in personalizing content, tracking customer spending, and automating the delivery of exclusive benefits. This ensures that your premium customers always feel valued and receive the best possible service without requiring significant manual effort.

ChatGPT Prompt to Help with Premium Service Tiers

> Hey ChatGPT, I need help designing a premium service tier system for my business. The goal is to offer exclusive benefits and content to our high-spending customers. Can you provide a step-by-step plan that includes:
> 1. Defining different tiers and their criteria.
> 2. Examples of exclusive benefits for each tier.
> 3. How to use AI to manage and personalize these tiers.
> 4. Strategies to promote the premium tiers to our customers.
> 5. Methods to measure the success and engagement of the premium service tiers.
> Thanks!

DAY 62: PAY-IN-FULL DISCOUNTS

A Pay-In-Full Discount is an incentive offered by businesses to encourage customers to pay the full amount of a product or service upfront, rather than opting for installment payments.

This type of discount can help businesses improve cash flow, reduce administrative costs associated with managing multiple payments, and provide financial stability.

For customers, it offers the advantage of receiving a discount or a better overall price by paying the total amount in one go.

ChatGPT Prompt to Help Implement Pay-In-Full Discounts

> I run a business and want to implement Pay-In-Full Discounts to encourage customers to pay the total amount upfront instead of in installments. I need assistance with the following:
> 1. Benefits and Considerations: Explain the benefits of offering Pay-In-Full Discounts for both my business and my customers. Also, discuss any potential drawbacks or considerations I should be aware of.
> 2. Setting Up Discounts: Provide step-by-step instructions on how to set up these discounts, including deciding on the discount percentage, communicating the offer to customers, and integrating it into our payment system.
> 3. Marketing the Offer: Suggest effective marketing strategies to promote the Pay-In-Full Discount to our target audience. Include tips on wording, channels to use, and how to highlight the value proposition.
> 4. Tracking and Adjusting: Advise on how to track the success of the Pay-In-Full Discount program and make necessary adjustments based on customer feedback and sales data.
> 5. Best Practices: Share best practices from other businesses that have successfully implemented Pay-In-Full Discounts.

DAY 63: SKIP-THE-LINE FEES

Skip-the-line fees allow customers to pay a premium for expedited service. This concept can be seen in various industries, such as theme parks offering fast passes, Instacart offering priority delivery, and FedEx providing overnight shipping.

By implementing a skip-the-line option, businesses can offer customers the convenience of faster delivery for an additional fee.

When I was young, we couldn't afford the fast pass at theme parks, so we had to wait in line for hours. Food was too expensive, so my mom packed sandwiches, and we had to walk all the way back to the parking lot to eat them.

Once I grew up and made a little money, I took my kids to Universal Studios. We stayed on the property and enjoyed early access, fast passes, and all the perks. It was worth paying the premium! Our wait time was almost non-existent, so we got to ride everything more than once.

This experience inspired us to offer same day rush fees at Zeus' Closet. Many customers are willing to pay 200-500% more, just for faster access. If you're not offering this, you're leaving money on the table.

ChatGPT Prompt:

> Help me develop a strategy to implement skip-the-line fees in my business. I want to offer customers the option to pay an additional fee for expedited service at checkout. How can I effectively communicate the benefits and value of this service to my customers, and what are some best practices for integrating this option seamlessly into my existing operations?

DAY 64: VIP SERVICE LEVELS

A VIP service level is a premium tier of customer service designed to provide enhanced support and exclusive offers to a select group of customers. This concept is similar to the first-class service on airlines, like Delta One, where passengers receive superior amenities and prioritized treatment. By creating a VIP customer segment, businesses can offer these top-tier customers special perks, priority support, and exclusive deals, thereby fostering loyalty and encouraging higher spending.

Key Benefits of VIP Service Levels

1. Enhanced Customer Support: Provide faster, more personalized responses and dedicated support channels.
2. Exclusive Offers: Offer special deals and discounts not available to regular customers.
3. Priority Treatment: Give VIPs priority in service queues, early access to new products, and more.

ChatGPT Prompt to Help Implement VIP Service Levels

> Hey ChatGPT, I want to create a VIP service level for my business that includes enhanced customer support and exclusive offers. Can you help me brainstorm?
>
> Then outline a detailed plan for this, including how to identify and segment VIP customers, what types of exclusive perks and priority support to offer, and how to effectively communicate these benefits to my VIP segment?
>
> Please include examples and best practices from leading companies to ensure a successful implementation.

DAY 65: PRODUCT BUNDLING

Product bundling is a marketing strategy where multiple products are packaged together and sold as a single combined unit, often at a discounted price compared to buying each item individually. This approach offers several benefits:

- Increased Perceived Value: Customers see more value in a bundle, feeling they get more for their money.
- Encourages Higher Spending: Bundles can entice customers to spend more than they initially intended.
- Inventory Management: Bundling helps move less popular products by pairing them with best-sellers.
- Customer Satisfaction: Customers appreciate the convenience of purchasing complementary products together.

Using AI for product bundling can streamline this process by analyzing customer behavior, purchase history, and product compatibility to suggest optimal bundles. This not only enhances customer experience but also boosts sales and profitability.

ChatGPT Prompt for Product Bundling

Here's a prompt you can use with ChatGPT to help you create effective product bundles:

> Hey ChatGPT, I need help creating product bundles for my online store. Could you assist me in developing a strategy using AI to automatically bundle complementary products? Here are the details:
> 1. Analyze Customer Data: Examine purchase history and customer behavior to identify products that are frequently bought together.
> 2. Suggest Complementary Products: Generate suggestions for product bundles that offer better value to customers.
> 3. Pricing Strategy: Recommend a discount strategy for these bundles to encourage bulk purchases while maintaining profitability.
> 4. Implementation Tips: Provide tips on how to present these bundles on my website to maximize visibility and appeal.
>
> Additionally, consider these specific product categories for bundling:
> - Category 1: [Insert product details]
> - Category 2: [Insert product details]
> - Category 3: [Insert product details]
>
> Thank you!

This prompt will guide ChatGPT to help you develop an AI-driven product bundling strategy that enhances customer value and drives higher sales.

DAY 66: EXTENDED WARRANTIES

Extended warranties are additional service plans offered by businesses to consumers, providing coverage beyond the standard warranty period. These warranties typically cover repairs or replacements for a set period after the manufacturer's warranty expires. They offer benefits to both businesses and consumers, creating a win-win situation.

Benefits for Businesses:

- Increased Average Order Value (AOV): Offering extended warranties can lead to higher sales as customers are often willing to pay extra for the added security, thereby increasing the overall transaction value.
- Customer Loyalty: Extended warranties can enhance customer satisfaction and loyalty, as consumers feel more secure about their purchases.
- Additional Revenue Stream: These warranties provide an extra source of income for the business without significant additional costs.

Benefits for Consumers:

- Peace of Mind: Consumers gain confidence knowing that their purchase is protected for a longer period, reducing the risk of unexpected repair costs.
- Convenience: Extended warranties often simplify the process of repairs or replacements, saving time and effort for consumers in case something goes wrong.

ChatGPT Prompt:

> Help me brainstorm and outline different types of extended warranties our business could offer. Additionally, draft some messaging that effectively communicates the benefits of these warranties to our customers, emphasizing peace of mind and convenience.

Example Output (yours may be different):

1. Brainstorming Different Types of Extended Warranties:
    - Standard Extended Warranty: Covers basic repairs and replacements beyond the manufacturer's warranty.
    - Comprehensive Protection Plan: Includes accidental damage, theft, and loss coverage.
    - Maintenance Plan: Offers regular maintenance services and check-ups.
    - Premium Support Package: Provides priority customer service and technical support.
    - Customization Warranty: Covers customized or personalized products, ensuring unique items are protected.
2. Draft Messaging:
    - Standard Extended Warranty: "Our Standard Extended Warranty offers extended protection for your purchase, ensuring repairs and replacements are covered beyond the manufacturer's warranty. This added layer of security increases the value of your investment and gives you peace of mind."
    - Comprehensive Protection Plan: "With our Comprehensive Protection Plan, you can rest easy knowing your product is safeguarded against accidental damage, theft, and loss. This plan not only enhances your shopping experience but also provides unparalleled convenience and security."
    - Maintenance Plan: "Keep your purchase in top condition with our Maintenance Plan, which includes regular check-ups and services. This plan not only extends the life of your product but also ensures it performs at its best, giving you added peace of mind."
    - Premium Support Package: "Our Premium Support Package offers priority customer service and technical support, ensuring you receive the assistance you need when you need it. Enjoy the convenience of fast and efficient help, making your experience with us even better."

DAY 67: GUARANTEES

Guarantees are promises or assurances provided by a business to its customers, indicating that certain conditions will be fulfilled. These can include satisfaction guarantees, product warranties, service commitments, and more. Guarantees help build trust and confidence in your brand, as they demonstrate your commitment to quality and customer satisfaction.

Why are Guarantees Important?

- Build Trust: Customers are more likely to purchase from a company that stands behind its products and services.
- Reduce Risk: Guarantees minimize the perceived risk for customers, making them more comfortable with their purchase decision.
- Differentiate Your Brand: A strong guarantee can set you apart from competitors.
- Encourage Repeat Business: Satisfied customers are more likely to return and recommend your business to others.

Types of Guarantees

- Satisfaction Guarantee: If a customer is not satisfied, they can get a refund or exchange.
- Money-Back Guarantee: If the product or service doesn't meet expectations, the customer gets their money back.
- Lifetime Guarantee: The product is guaranteed for the lifetime of the original owner.
- Service Guarantee: Assurance of a specific level of service quality.

ChatGPT Prompt to Help with Crafting a Guarantee

> Hey ChatGPT, I need help crafting a strong guarantee for my business that will build trust with my customers and set us apart from competitors. We sell [product/services], and I want the guarantee to emphasize our commitment to quality and customer satisfaction. Can you help me create a compelling guarantee statement? Also, please provide a few different types of guarantees that might work well for my business.

DAY 68: ORDER BUMPS

Order Bumps are additional items or offers that customers can easily add to their cart during the checkout process by simply checking a box. These are designed to be enticing and relevant to the customer's primary purchase, similar to how magazines and candy bars are positioned near the checkout lanes in grocery stores to encourage impulse buys.

Order bumps can help increase the average order value and enhance the overall customer experience by providing convenient, complementary products.

Examples of Order Bumps

- Fitness and Health Store:
    - Protein Shaker Bottle: Add a high-quality shaker bottle to your order.
    - Fitness Bands: Enhance your workout with a set of resistance bands.
    - Nutrition Guide: Get a comprehensive nutrition guide to complement your fitness journey.
- Tech Gadgets Store:
    - Extended Warranty: Protect your gadget with an extended warranty.
    - Screen Protectors: Add a set of screen protectors to keep your device safe.
    - Portable Charger: Stay powered up with a compact portable charger.
- Beauty and Skincare Store:
    - Travel-Sized Products: Try travel-sized versions of our best-sellers.
    - Makeup Brushes: Add a set of professional makeup brushes.
    - Face Masks: Enhance your skincare routine with a pack of nourishing face masks.
- Home Decor Store:
    - Candle Set: Add a set of scented candles to create a cozy atmosphere.
    - Decorative Throw Pillows: Complement your new decor with matching throw pillows.
    - Wall Art: Enhance your space with a piece of stylish wall art.
- Pet Supplies Store:
    - Pet Treats: Add a bag of delicious and healthy treats for your pet.
    - Pet Toys: Include a fun toy to keep your pet entertained.

- Grooming Kit: Maintain your pet's coat with a grooming kit.
- Bookstore:
    - Bookmarks: Add a set of themed bookmarks to your order.
    - Reading Light: Get a clip-on reading light for late-night reading.
    - Book Sleeve: Protect your books with a custom book sleeve.
- Outdoor and Adventure Store:
    - Water Bottle: Add a durable water bottle to stay hydrated on your adventures.
    - Multi-Tool: Equip yourself with a handy multi-tool.
    - First Aid Kit: Be prepared with a compact first aid kit.
- Fashion Boutique:
    - Jewelry: Add a stylish piece of jewelry to complete your look.
    - Socks or Tights: Complement your outfit with a pair of socks or tights.
    - Scarf or Hat: Accessorize with a matching scarf or hat.
- Cooking and Kitchen Store:
    - Spice Mixes: Enhance your dishes with a set of gourmet spice mixes.
    - Cookbook: Add a cookbook with recipes that complement your new kitchen gadget.
    - Kitchen Utensils: Include a set of high-quality kitchen utensils.

These examples showcase how order bumps can be tailored to different types of online stores, offering products that naturally complement the main purchase and encourage impulse buys.

ChatGPT Prompt to Help Implement Order Bumps for Your Business

> I run an online store and I'm looking to implement order bumps to increase my average order value. Order bumps are items that customers can add to their cart with a simple checkbox during checkout, similar to impulse buys like magazines and candy bars at the grocery store. Please help me with the following:
> 1. Identify Potential Order Bumps:
>   - Suggest complementary items for my main products (e.g., if my main product is a custom Greek jacket, what smaller items could be offered as order bumps?).
> 2. Craft Compelling Descriptions:
>   - Write engaging and persuasive descriptions for these order bumps to entice customers to add them to their cart.
> 3. Placement and Strategy:
>   - Provide tips on the optimal placement of these order bumps during the checkout process.
>   - Offer strategies to test and optimize the effectiveness of these order bumps.
> 4. Examples:
>   - Give specific examples of successful order bumps used by other e-commerce businesses for inspiration.
> 5. Best Practices:
>   - Outline best practices for ensuring these order bumps do not disrupt the customer's checkout experience.

Using this prompt, ChatGPT can help you brainstorm and implement effective order bumps tailored to your online store, maximizing both customer satisfaction and sales.

DAY 69: HIGH-TICKET UPSELLS

High-ticket upsells refer to the strategy of offering premium, higher-priced products or services to customers who have already made a purchase or are in the process of purchasing.

This approach aims to maximize the value of each customer transaction by encouraging them to invest in more expensive offerings, which often come with enhanced features, benefits, or exclusive access.

Why Use High-Ticket Upsells?

- Increased Revenue: By offering high-ticket items, you can significantly boost your overall sales and profitability.
- Enhanced Customer Experience: Premium products or services often provide added value, improving customer satisfaction and loyalty.
- Maximize Customer Lifetime Value: Encouraging customers to invest more can lead to longer-term relationships and repeat business.
- Efficient Marketing: Targeting existing customers who already trust your brand can be more cost-effective than acquiring new customers.

Examples of High-Ticket Upsells

- Offering a premium version of a product with additional features.
- Providing exclusive access to one-on-one coaching or consulting services.
- Selling advanced training programs or courses.
- Upselling luxury or personalized items.

ChatGPT Prompt to Help with High-Ticket Upsells

> Hey ChatGPT, I need help creating a high-ticket upsell strategy for my business. Here's some information about my current offerings and customer base:
> 1. Current Products/Services:
>   - [Briefly describe your main products or services]
> 2. Customer Profile:
>   - [Describe your typical customer demographics, preferences, and buying behavior]
> 3. Goals for Upselling:
>   - [State your revenue goals, customer experience objectives, etc.]
>
> Could you suggest some high-ticket upsell ideas, and provide tips on how to effectively present these offers to my customers? Additionally, I'd appreciate any advice on crafting compelling marketing messages for these upsells. Thanks!

Feel free to tweak the prompt with specific details about your business to get tailored advice from ChatGPT.

DAY 70: AI-GENERATED PRODUCT DESCRIPTIONS

AI-generated product descriptions involve using artificial intelligence to create detailed, persuasive narratives about products. These descriptions emphasize the value and benefits, aiming to attract and convince potential customers. The goal is to highlight the unique features and advantages of premium products, making them appealing and desirable to the target audience.

Benefits:

- Consistency: Ensures uniformity in tone and style across all product descriptions.
- Efficiency: Saves time by automating the content creation process.
- Persuasiveness: Leverages AI's ability to craft compelling narratives that emphasize product benefits.
- SEO Optimization: Can include keywords to improve search engine rankings and visibility.

ChatGPT Prompt

To generate these product descriptions using ChatGPT, you can use the following prompt:

> Create a detailed and persuasive product description for our [product name], a premium [category] item. Highlight the key features, benefits, and unique selling points. Emphasize how this product adds value to the customer's life, solving their specific problems or enhancing their experience. Make the description engaging, informative, and compelling. Include any technical specifications and real-world applications to demonstrate its practicality and superiority over competitors. Ensure the tone aligns with our brand's voice.

Assuming you have your custom instructions tailored and turned on, ChatGPT should already know your brand's voice. If not, describe it for ChatGPT at the end of the prompt above.

Example:

> Create a detailed and persuasive product description for our UltraComfort Memory Foam Mattress, a premium sleep solution. Highlight the key features, benefits, and unique selling points. Emphasize how this mattress adds value to the customer's life by providing unparalleled comfort and support, ensuring a restful night's sleep. Make the description engaging, informative, and compelling. Include any technical specifications, such as the type of memory foam used, and real-world applications, like its suitability for people with back pain. Ensure the tone aligns with our brand's voice, which is professional and luxurious.

DAY 71: AUTOMATED CUSTOMER FEEDBACK

What if you had a system that collects and analyzes feedback from customers to help improve your products and services? This system uses various channels such as emails, surveys, social media, and in-app messages to gather feedback automatically. The data is then analyzed using AI and machine learning to identify trends, sentiments, and areas of improvement. This approach enables businesses to respond quickly to customer needs, enhance customer satisfaction, and continuously refine their offerings based on real-time insights.

ChatGPT Prompt to Help with Automated Customer Feedback

> Hey ChatGPT, I need help setting up an automated customer feedback system. Here's what I need:
> 1. Collection Methods: Recommend the best channels for collecting customer feedback (e.g., emails, surveys, social media, in-app messages).
> 2. Tools and Software: Suggest tools and software that can help automate the collection and analysis of customer feedback.
> 3. Data Analysis: Provide guidance on how to analyze the collected feedback using AI and machine learning to identify trends and sentiments.
> 4. Actionable Insights: Advise on how to translate the analyzed data into actionable insights to improve our products and services.
> 5. Best Practices: Share best practices for maintaining a continuous feedback loop and ensuring high response rates from customers.
>
> Thank you!

Tools to Increase Repeat Purchase Frequency and Retention

Boosting repeat purchase frequency and retaining your customers is the secret sauce to long-term business success. It's not just about making that first sale; it's about creating lasting relationships that keep customers coming back for more.

With the power of AI and automation at your fingertips, you can design personalized and engaging experiences that encourage your customers to return and spend more. These tools allow you to understand your customers' needs and habits on a deeper level, making it easier to deliver exactly what they want—when they want it.

I challenge you to take one day to learn and try each step in this chapter. You never know what's possible until you give it a shot!

Remember *The Rule of 26%*? If the average client shops with you four times a year, could you encourage them to shop just one more time? That extra purchase could significantly impact your bottom line. So, dive into these actionable ideas to increase repeat purchase frequency using AI and automation, and watch your business thrive like never before.

DAY 72: INTERACTIVE AND GAMIFIED CONTENT

Incorporating interactive and gamified content into your website can revolutionize the way you engage with your audience. By leveraging elements of game design, you can create a more engaging, enjoyable, and immersive experience for your visitors. This not only keeps users on your site longer but also drives higher conversion rates as they interact more deeply with your content.

Types of Interactive Content

- Quizzes and Surveys: Quizzes and surveys are excellent tools for engaging users while gathering valuable insights about their preferences and behaviors. They can be fun and informative, encouraging users to share their results and experiences.
- Interactive Infographics: Interactive infographics transform static data into dynamic experiences. Users can explore information at their own pace, clicking through different sections to learn more.
- Polls and Voting Systems: Polls and voting systems encourage user participation by allowing them to voice their opinions. This not only increases engagement but also provides you with direct feedback from your audience.
- Interactive Calculators and Tools: Tools like budget calculators, fitness trackers, and personalized recommendation engines provide value by helping users solve specific problems or achieve certain goals.

Gamification Elements

- Points and Rewards Systems: Implement a points system where users earn rewards for various actions, such as engaging with content, making purchases, or participating in community activities.
- Leaderboards: Leaderboards create a sense of competition and achievement. They motivate users to engage more by showing their progress relative to others.
- Badges and Achievements: Badges and achievements recognize and reward users for reaching milestones or completing specific tasks, fostering a sense of accomplishment.

- Challenges and Missions: Create challenges and missions that users can complete for rewards. These can be related to your products or services, or even broader community activities.

Implementation Tips

- Personalize Experiences: Use AI to tailor interactive and gamified experiences to individual users based on their preferences and behaviors.
- Seamless Integration: Ensure that gamification elements are smoothly integrated into your website, enhancing the user experience without being intrusive.
- Track and Analyze Data: Continuously monitor user engagement and interaction data to refine your strategy and improve the effectiveness of your gamified content.

One effective example of gamification is my own 6-Pack Dads fitness program. In this program, participants receive points for engaging with the content, correctly answering fitness-related questions, and turning in their workouts, weigh-ins, and food journals. This points system creates a fun, competitive environment that motivates participants to stay committed to their fitness goals. As a result, we see higher engagement levels and improved outcomes for the participants.

Tools and Resources

- Gamify: A platform that helps you integrate gamification elements into your website.
- Outgrow: A tool for creating interactive quizzes, calculators, and polls.
- Typeform: A user-friendly tool for building engaging surveys and quizzes.
- Playbuzz: A platform for creating and sharing interactive content.

Example Prompt for ChatGPT

> Act as a gamification expert. I need help creating interactive and gamified content for my website to boost user engagement and conversions. Can you provide an outline and tips for implementing quizzes, surveys, interactive infographics, polls, and gamification elements like points systems, leaderboards, badges, and challenges? Include examples of successful case studies and tools/resources for implementation.

By introducing interactive and gamified content to your business, you can significantly enhance user interaction, improve engagement, and drive business growth. This approach not only makes your content more enjoyable but also fosters a deeper connection with your audience.

DAY 73: APPLY FOR GRANTS

Imagine receiving a financial boost that doesn't need to be repaid. That's what a grant offers—a sum of money given by entities like governments, corporations, foundations, or trusts to support specific projects or initiatives. Grants are the lifeblood for many businesses, non-profits, and individuals, fueling innovations and impactful work without the burden of repayment.

The idea of grants isn't new. Think back to the age of patronage, where artists, scholars, and explorers were funded by wealthy benefactors. Fast forward to today, and this concept has evolved. Modern grants are more structured and targeted, provided by organizations aiming to drive progress in areas like science, education, and community development.

Why It Works

Grants are powerful because they unlock potential. They enable projects that might otherwise struggle to find funding. For grantors, it's a chance to see their mission and values materialize through the work they support. For recipients, it's an opportunity to focus on creating impact without the financial stress of loans.

How to Apply for Grants Using AI

This is where technology, particularly AI, can transform your grant application process. Here's how ChatGPT can help:

- Researching Opportunities: Let AI do the heavy lifting by finding grants that match your project's goals. Provide ChatGPT with your project details, and watch as it sifts through countless opportunities to find the best fits. For U.S. government grants, Grants.gov is your go-to resource. It offers a comprehensive database of available federal grants.
- Writing Proposals: Crafting a compelling proposal is an art. ChatGPT can help you draft, edit, and refine your proposals, ensuring they are clear, coherent, and persuasive. Simply input your project's key information, and get a well-structured draft ready to impress.

- Proofreading and Editing: Precision is key in grant applications. Use ChatGPT to review your proposals, catching grammatical errors and ensuring a polished final product.
- Generating Supporting Documents: Need a budget, timeline, or letter of support? ChatGPT can help create these documents, ensuring they are detailed and professional.
- Simulating Reviews: Want to know how your application might be received? ChatGPT can simulate the review process, offering feedback from a grant evaluator's perspective. This helps you identify and fix potential weaknesses before submission.

Some of the Top Tools for Applying for Grants

- Grants.gov: The official source for U.S. federal grants, providing a comprehensive database and application submission tools.
- GrantStation: Find comprehensive grant opportunities and resources.
- Instrumentl: Discover, track, and manage grants efficiently.
- GrantWatch: A specialized search engine for nonprofits and small businesses.
- ChatGPT: Your go-to tool for drafting, editing, and refining grant proposals.

Little-Known Tips and Tricks

- Tailor Each Application: Customize your proposal to align with the specific goals and criteria of each grant.
- Show Impact: Clearly articulate the potential impact of your project and how it aligns with the grantor's mission.
- Seek Feedback: Get input from colleagues or mentors to strengthen your proposal.
- Follow Up: If you don't get the grant, ask for feedback to improve future applications.

Action Item

Ready to get started? Try this prompt with ChatGPT:

Act as a professional grant writer who has a winning streak for small business grants. I am planning to apply for a grant to fund a project focused on [insert project details]. Can you help me draft an outline for the proposal, including the key sections I should cover and tips for making my application stand out?

DAY 74: SPECIAL OFFERS AND COUPONS

Creating automated special offers or coupons is an effective strategy to encourage repeat purchases from existing clients. By offering unexpected VIP perks or personalized discount codes, you can re-engage customers who haven't made a purchase in a while, making them feel valued and appreciated.

Example: Say you want to send personalized discount codes to customers who haven't made a purchase in the last three months.

ChatGPT Prompt to Help with This Task:

> Hi ChatGPT,
>
> I need assistance in creating a strategy for automated special offers and coupons to encourage repeat purchases from my existing clients. Here are the specifics:
>
> 1. Objective: Encourage repeat purchases from customers who haven't made a purchase in the last three months.
> 2. Target Audience: Existing clients who have not made a purchase in the last three months.
> 3. Offer Details: Personalized discount codes and unexpected VIP perks.
> 4. Execution Plan:
>   - Identify the target customers using our CRM system.
>   - Generate personalized discount codes.
>   - Create automated email templates to send out these offers.
>   - Monitor the response and effectiveness of the campaign.
>
> Could you provide a detailed plan and example email templates for this campaign? Also, any tips on making these offers more appealing would be greatly appreciated. Thank you!

DAY 75: SUBSCRIPTIONS

Implementing AI-driven membership or subscription services is a strategic approach to providing regular value and convenience to your customers. This model can enhance customer loyalty and create a steady revenue stream. By automating billing and delivery schedules, you ensure a seamless experience for your subscribers. To gain insights into different subscription business models, John Warrilow's book *The Automatic Customer* is an excellent resource. It outlines nine types of subscription models that businesses can leverage to create consistent revenue and deepen customer relationships.

- Membership Website Model: Provides access to exclusive content, resources, or community features.
- All-You-Can-Eat Library Model: Offers unlimited access to a library of content or products for a recurring fee.
- Private Club Model: Grants members access to exclusive events, products, or services.
- Front-of-the-Line Model: Provides priority access to new products, services, or limited availability items.
- Consumables Model: Regularly delivers consumable products (e.g., food, toiletries) to subscribers.
- Surprise Box Model: Sends curated boxes of products tailored to the subscriber's interests on a regular basis.
- Simplifier Model: Takes over a task or responsibility, simplifying the customer's life (e.g., meal kits, house cleaning).
- Network Model: Creates value by connecting subscribers with a network or community (e.g., professional networks, social platforms).
- Peace-of-Mind Model: Provides services or products that give subscribers a sense of security and assurance (e.g., insurance, maintenance plans).

ChatGPT Prompt To Help You Design Subscription Models

I'm considering implementing AI-driven membership or subscription services for my business. Could you provide a detailed explanation of how this model works and how it can benefit my company?

Additionally, can you summarize the nine subscription business models described in John Warrilow's *The Automatic Customer* and suggest which models might be most applicable to my business?

DAY 76: REGULAR LEARNING CONTENT

The goal is to consistently deliver valuable learning content through email and social media to keep customers engaged and informed. This involves creating an email sequence that sends out weekly tips and tutorials related to your products and building an online learning portal. This approach not only helps customers get the most out of your products but also fosters a sense of community and ongoing education.

I once stayed subscribed to a certain marketing software solely because of the learning content and community that came with it. Despite the software being buggy and not particularly impressive, the educational resources and the sense of belonging kept me engaged and loyal. You can build the same type of platform for your business.

ChatGPT Prompt

> I want to develop a strategy for delivering regular learning content to my customers via email and social media. Here's what I need help with:
>
> 1. Creating an email sequence that sends out weekly tips and tutorials related to our products. Each email should be engaging, informative, and provide actionable insights that customers can use immediately.
>
> 2. Building an online learning portal where customers can access a library of tutorials, guides, and other educational resources. This portal should be easy to navigate, visually appealing, and regularly updated with new content.
>
> 3. Developing a content calendar for both email and social media to ensure consistent and timely delivery of learning materials. This calendar should include a mix of tutorials, tips, customer success stories, and interactive content like quizzes or challenges.
>
> 4. Incorporating community-building elements into the learning content to foster a sense of connection and loyalty among customers.
>
> Can you provide detailed steps and suggestions for each of these components to ensure the learning content is engaging and effective? I'd like to leverage AI & automation for this as much as possible.

DAY 77: 100-DAY ONBOARDING PLAN

The 100-Day Onboarding Plan is a strategic approach to welcoming and integrating new customers into your business using Joey Coleman's "First 100 Days" methodology. The goal is to create remarkable customer experiences that ensure customer retention and loyalty. This plan involves guiding customers through eight phases of the customer lifecycle, using a variety of communication tools to enhance their experience at each phase.

Definitely read Joey Coleman's book *Never Lose a Customer Again: Turn Any Sale into Lifelong Loyalty in 100 Days* to learn more.

The phases include: Assess, Admit, Affirm, Activate, Acclimate, Accomplish, Adopt, and Advocate.

The plan typically involves a series of structured interactions, including emails, phone calls, videos, physical mail, and in-person meetings. By carefully managing these touchpoints, you can build a strong relationship with your customers, reduce buyer's remorse, and turn them into advocates for your brand.

Try the ChatGPT prompt on the next page to help develop a 100-day onboarding sequence for your business.

You are a highly skilled customer experience strategist. Your task is to help develop a 100-day onboarding email sequence using Joey Coleman's First 100 Days methodology. The goal is to create a remarkable customer experience for new customers of [Your Business Name]. The sequence should welcome new customers and guide them through our products and services, ensuring they feel valued and engaged throughout the onboarding process.

1. **Assess Phase (Pre-Day 1)**
   - **Email 1:** Welcome email introducing our brand, mission, and the benefits of our products/services. Provide educational content and free resources to help prospects evaluate our offerings.
   - **Email 2:** Share customer testimonials and case studies to build trust and set expectations.

2. **Admit Phase (Day 1 - Purchase)**
   - **Email 3:** Personalized thank you email for their purchase, highlighting the excitement and possibilities of their new journey with us.
   - **Email 4:** Send a welcome video from the team expressing gratitude and enthusiasm.

3. **Affirm Phase (Days 2-14)**
   - **Email 5:** Address potential buyer's remorse with a reassuring message and share additional testimonials.
   - **Email 6:** Provide a "Keep the Faith" video reaffirming their purchase decision and highlighting key benefits.

4. **Activate Phase (Days 15-30)**
   - **Email 7:** Confirm receipt of the product/service with tips on getting started and maximizing its value.
   - **Email 8:** Share a personalized video with success stories and usage tips from other customers.

5. **Acclimate Phase (Days 31-60)**

- **Email 9:** Provide advanced tips and tricks for using the product/service, along with a quick survey to gather feedback.
   - **Email 10:** Offer a group call or teleseminar for personalized coaching and support.

6. **Accomplish Phase (Days 61-80)**
   - **Email 11:** Congratulate the customer on reaching a milestone and share a celebratory video from the team.
   - **Email 12:** Highlight the benefits they've experienced so far and remind them of the initial goals.

7. **Adopt Phase (Days 81-90)**
   - **Email 13:** Provide an "Expert User" guide with advanced usage tips and success stories from long-term customers.
   - **Email 14:** Conduct an in-depth success survey to gather insights and personalize future interactions.

8. **Advocate Phase (Days 91-100)**
   - **Email 15:** Ask for a testimonial or referral, offering an incentive for sharing their positive experience.
   - **Email 16:** Send a final thank you email with a personalized gift or offer, and invite them to join a loyalty program or community.

Please create detailed content for each email, ensuring they are engaging, personalized, and aligned with our brand voice. The emails should also include calls-to-action to encourage customer interaction and feedback.

By following this prompt, you can create a comprehensive and engaging 100-day onboarding email sequence that helps new customers feel welcomed, valued, and supported throughout their journey with your business.

DAY 78: AFTER-PURCHASE PRODUCT SUGGESTION EMAILS

After-purchase product suggestion emails are a strategic way to engage with customers following their purchase. These emails serve multiple purposes:

- Enhance Customer Experience: By checking in and ensuring satisfaction with the initial purchase.
- Increase Sales: By suggesting complementary or related products based on their previous purchase.
- Build Loyalty: By showing personalized attention and care, fostering a deeper connection with your brand.

By leveraging AI and automation, you can create personalized, timely, and relevant follow-up emails without manual effort. AI can analyze purchase data to recommend the best products for each customer, while automation tools schedule and send these emails at optimal times.

ChatGPT Prompt to Help Craft These Emails

> ChatGPT, I need help crafting after-purchase product suggestion emails for my business. Here are the details:
> 1. Business Name: [Your Business Name]
> 2. Customer Purchase Data: [Details about recent purchases, e.g., product name, category]
> 3. Suggested Products: [List of complementary products you want to suggest]
> 4. Tone: [Friendly, professional, casual, etc.]
> 5. Frequency: [How often to send these emails]
>
> Please create a series of engaging follow-up emails that check in with customers about their recent purchase and suggest related products they might be interested in. Include subject lines and call-to-action prompts. Also, provide a brief schedule for when these emails should be sent.

DAY 79: AI-DRIVEN LOYALTY PROGRAMS

Loyalty programs are marketing strategies designed to encourage customers to continue shopping at or using the services of a business associated with the program. These programs reward customers for their repeat purchases and high spending, fostering customer retention and increasing customer lifetime value. Common rewards include discounts, points redeemable for products or services, exclusive offers, and early access to new products.

Benefits of Loyalty Programs

- Increased Customer Retention: Encourages customers to return for repeat purchases.
- Higher Customer Lifetime Value: Boosts the overall revenue generated from a single customer over time.
- Enhanced Customer Experience: Provides customers with incentives and rewards, improving their overall shopping experience.
- Data Collection: Offers insights into customer preferences and buying behavior, which can be used to personalize marketing efforts.

Steps to Implement a Loyalty Program

1. Define Objectives: Determine what you want to achieve with your loyalty program (e.g., increased sales, customer retention).
2. Choose the Type of Program: Decide on the structure of your program (e.g., points-based, tiered rewards, paid memberships).
3. Select Rewards: Choose attractive and attainable rewards that align with your customers' interests.
4. Integrate Technology: Use software or apps to manage and track the program.
5. Promote the Program: Market the loyalty program through various channels to ensure customer awareness.
6. Monitor and Adjust: Regularly review the program's performance and make adjustments as needed.

ChatGPT Prompt to Help with Loyalty Programs

Hi ChatGPT, I need help creating a loyalty program for my business. Could you guide me through the process? Specifically, I need assistance with:

1. Defining the objectives of my loyalty program.
2. Choosing the best type of program for my business.
3. Selecting appropriate rewards that will appeal to my customers.
4. Recommendations for software or apps to manage the loyalty program.
5. Effective ways to promote the program to ensure maximum customer engagement.
6. Tips on monitoring the program's performance and making necessary adjustments.

Any detailed suggestions or examples you can provide would be greatly appreciated!

Follow up with:

Could you guide me through how AI can assist with the following aspects?

1. Analyzing customer data to gain insights and segment customers.
2. Personalizing rewards and offers based on customer behavior.
3. Automating communication and program management tasks.
4. Detecting and preventing fraud within the loyalty program.
5. Enhancing the overall customer experience with AI tools.
6. Making data-driven decisions to improve program performance.

Please provide detailed suggestions and examples of AI tools or techniques that can be used for each aspect.

DAY 80: HIRING RIGHT

The hiring process is a series of steps that companies follow to attract, select, and onboard new employees. It typically includes creating job descriptions, formulating interview questions, drafting offer letter templates, and establishing a comprehensive onboarding process.

ChatGPT Prompts for Assistance

You can use the following prompt to get ChatGPT's help with your hiring process:

> ChatGPT, I need assistance with the hiring process for my company. Please help me with the following:
>
> 1. **Job Descriptions**: Create detailed job descriptions for [specific roles, e.g., Marketing Manager, Software Developer] that include responsibilities, required qualifications, and desired skills.
> 2. **Interview Questions**: Provide a list of effective Topgrading-style interview questions tailored to assess candidates for these roles.
> 3. **Offer Letter Templates**: Draft offer letter templates for the roles, including sections for salary, benefits, start date, and any other important terms.
> 4. **Onboarding Process**: Outline a comprehensive onboarding process for new hires, including orientation, training programs, and integration into the company culture.
>
> Thank you!

Feel free to customize the roles and details according to your specific needs.

Here are some suggested follow up prompts:

What strategies can we employ to ensure our job descriptions are more inclusive and inviting to candidates from diverse backgrounds?

Offer suggestions and provide language examples that can help us create an inclusive job description for a [job title] position at [company name].

I own a [size] business in the [industry] industry. We are currently automating our hiring process and are looking for recommendations on HR software suites. Search the web for the most up-to-date options and suggest three HR software suites.

DAY 81: LEADERSHIP

As a CEO, one of your key roles is to communicate your vision to your team in a way that inspires and motivates them to be part of it. Effective communication of vision is crucial for aligning your team, fostering a sense of purpose, and driving collective efforts toward common goals.

However, many business leaders make mistakes in this process, which can lead to confusion, disengagement, and lack of alignment within the organization.

Common Mistakes and How to Avoid Them

- Lack of Clarity:
    - Mistake: Vague or overly complex visions that are hard to understand.
    - Solution: Clearly articulate a concise and compelling vision that everyone can understand and remember.
- Inconsistent Messaging:
    - Mistake: Changing the vision frequently or communicating different messages to different groups.
    - Solution: Ensure consistency in your message across all levels of the organization. Reinforce the vision regularly through various communication channels.
- Failure to Connect Emotionally:
    - Mistake: Presenting the vision in a dry, uninspiring manner.
    - Solution: Use storytelling and real-life examples to connect emotionally with your team. Highlight how the vision aligns with their values and aspirations.
- Lack of Engagement:
    - Mistake: Communicating the vision as a top-down directive without involving the team.
    - Solution: Encourage feedback and participation from your team. Make them feel like co-creators of the vision.

- Ignoring Individual Contributions:
    - Mistake: Failing to show how each team member's role contributes to the vision.
    - Solution: Clearly outline how individual and team efforts align with and drive the vision forward. Recognize and celebrate contributions.

ChatGPT Prompt to Help Communicate Your Vision

> I'm looking for a comprehensive guide to help me, as a CEO, effectively communicate my vision to my team in a way that inspires and motivates them to be part of it. The guide should include:
>
> 1. Steps to clearly articulate a concise and compelling vision.
> 2. Strategies to ensure consistent messaging across the organization.
> 3. Techniques to connect emotionally with the team through storytelling and real-life examples.
> 4. Methods to engage the team and make them feel like co-creators of the vision.
> 5. Ways to highlight how individual contributions align with and drive the vision forward.
>
> Additionally, please include common mistakes business leaders make in this process and how to avoid them. Provide practical tips and examples to illustrate each point.

This prompt will guide ChatGPT in creating a detailed and practical guide for you, ensuring that you communicate your vision effectively and avoid common pitfalls.

DAY 82: CORE VALUES

Core Values are the fundamental beliefs and guiding principles that dictate the behavior and decision-making process within an organization. They shape the company's culture, influence how employees interact with each other and with clients, and establish the company's identity both internally and externally. Core values are critical for:

- Defining Company Culture: They set the tone for the workplace environment.
- Guiding Decisions: They help employees make decisions that align with the company's mission and vision.
- Building Brand Identity: They contribute to the company's public image and reputation.
- Enhancing Employee Alignment: They ensure that all team members are working towards the same goals and values.

ChatGPT Prompts to Help Develop Company Core Values

> Hey ChatGPT, I need to develop a set of core values for my company. Can you help me create a comprehensive list of core values that reflect our mission and vision? Here's some context about our company:
> 1. Mission: [Insert your company's mission statement here]
> 2. Vision: [Insert your company's vision statement here]
> 3. Industry: [Describe the industry your company operates in]
> 4. Culture: [Briefly describe the desired company culture]
> 5. Key Principles: [List any key principles or philosophies your company adheres to]
> Based on this information, can you suggest a list of core values with brief descriptions for each?

Here's another approach:

> Help me identify my company's core values. I will provide you with positive reviews we've received from customers, and a list of qualities that our team's top A-players all possess. After reading this information, reply with 3-5 single-word core values. Even better if it is an acronym.

Way before ChatGPT existed, we actually used this method—as a team, manually, with a whiteboard—to determine our Core Values.

My company's Core Values are:

- Promises
- Accuracy
- Logic
- Lifelong Learning
- Systems

Search YouTube for "Zeus' Closet core values" to see a video of them in action.

You may want to create a core values video for your company too! Just continue the conversation and ask ChatGPT to help you write the script.

> ChatGPT, I want to make a short movie about our company's core values. They are [list them here]. Please write a script for a 3-minute video.

DAY 83: PERFORMANCE REVIEWS

Employee performance reviews are a systematic way of evaluating job performance and overall contribution to the organization. These reviews help in understanding individual achievements, areas of improvement, and setting future goals. They provide a structured opportunity for feedback, professional development, and alignment with the company's objectives.

Typically, these reviews include assessing skills, accomplishments, teamwork, communication, and goal attainment.

When your employees perform better, they deliver a better experience to your clients, which improves retention and will likely increase repeat business.

Key Components of Performance Reviews

1. Goals and Objectives: Reviewing if team members met their set goals.
2. Strengths and Achievements: Highlighting key accomplishments and strengths.
3. Areas for Improvement: Identifying areas where the team member can grow.
4. Feedback and Coaching: Providing constructive feedback and guidance.
5. Future Goals: Setting new goals and expectations for the upcoming period.

Tip: Evaluate your team members' performance based on your company's Core Values, and make the criteria clear and transparent.

ChatGPT Prompts to Assist with Team Member Performance Reviews

> ChatGPT, I need help preparing performance reviews for my team members. Please guide me through creating a detailed and effective performance review. Include sections for goals and objectives, strengths and achievements, areas for improvement, feedback and coaching, and future goals. Provide me with examples of questions to ask in each section and tips on delivering constructive feedback.

> Act as an HR expert. I am writing a performance review about one of my [department] team members.
>
> Here are some areas where the team member performs well:
> [list positives here]
>
> Here are some areas where there is room for improvement:
> [list areas of improvement here]
>
> Help me write a performance review that is encouraging, but also makes sure [he/she] will understand the mistakes and avoid them in the future. Include any necessary legal language too.

Be sure to cross reference ChatGPT's output with employment laws. ChatGPT can make mistakes.

DAY 84: HANDWRITTEN THANK YOU NOTES

Handwritten cards stand out in an age dominated by digital communication. Receiving a tangible, hand-penned note creates a memorable experience, making the recipient feel special. This personal touch fosters stronger connections and loyalty. Studies have shown that personalized gestures, such as handwritten notes, significantly enhance customer satisfaction and retention. For businesses, this translates to increased customer loyalty and repeat business.

Leveraging AI can streamline the process of writing handwritten cards without losing the personal touch. Services like Handwrytten allow businesses to automate the creation and sending of handwritten notes. You can integrate this with customer relationship management (CRM) tools to trigger the sending of a card after specific customer interactions, such as making a purchase or attending an event.

By using platforms like Handwrytten in conjunction with Zapier, you can create workflows that automatically send a card when a certain condition is met in your CRM. For example, you can set up a Zapier integration to send a thank you card whenever a new client signs up or after a customer makes their fifth purchase.

The Best Tools for Handwritten Cards

- Handwrytten: Offers customizable, automated handwritten notes with a wide selection of handwriting styles and card designs.
- Zapier: Connects Handwrytten with your CRM and other tools to automate the sending process.
- HubSpot: A CRM that can be integrated with Handwrytten via Zapier for seamless customer interactions.
- Salesforce: Another CRM option that works well with Handwrytten for larger businesses looking to maintain personalized customer relations at scale.
- Thankster: An alternative service providing personalized handwritten notes.

Little Known Tips and Tricks

- Personalization: Always include the recipient's name and a specific detail about your interaction to make the card feel truly personalized.
- Timing: Send handwritten cards at unexpected times, not just during holidays or after transactions. For example, a card on the anniversary of their first purchase with you can be a delightful surprise.
- Handwriting Style: Choose a handwriting style that closely matches your own or reflects your brand's personality.

Action Item

To get started with automating your handwritten cards, set up an account with Handwrytten and connect it to your CRM via Zapier. Use the following ChatGPT prompt to brainstorm unique card messages tailored to different customer interactions:

> Generate ten unique handwritten thank you card messages for new clients, each including a personalized touch related to their first purchase.
>
> Here's some information about client #1 [name, product/service purchased]

DAY 85: BIRTHDAY AND ANNIVERSARY OFFERS

Birthday and anniversary offers are personalized promotions sent to customers to celebrate their special days. By automating these offers, you can enhance customer experience, make them feel valued, and boost loyalty. This strategy not only shows customers that you care about them but also encourages repeat purchases, driving increased sales and customer retention.

ChatGPT Prompt:

> Hey ChatGPT, I need help crafting a campaign for automated birthday and anniversary offers for my customers. Can you assist me with the following?
> 1. Email Templates:
>   - A warm and personalized birthday email template.
>   - An engaging anniversary email template.
> 2. Offer Ideas:
>   - Discounts, freebies, or special gifts for birthdays and anniversaries.
>   - Creative ways to present these offers to make customers feel special.
> 3. Timing and Scheduling:
>   - Best practices for when to send these emails.
>   - Tips for setting up automated scheduling.
> 4. Personalization Tips:
>   - Ways to customize these emails to each customer.
>   - Ideas for adding a personal touch that will resonate with recipients.
> 5. Follow-up Strategy:
>   - How to follow up after the special day to thank customers and remind them of their offers.
>   - Ideas for post-purchase engagement to encourage loyalty.
>
> Please provide detailed suggestions and examples for each of these points to help me create an effective and heartfelt campaign. Thanks!

DAY 86: FEEDBACK AND IMPROVEMENT SURVEYS

Feedback and Improvement Surveys are a crucial tool for gathering customer insights to enhance your products and services. By automating these surveys, you can efficiently collect valuable feedback from your customers.

One popular metric used in these surveys is the Net Promoter Score (NPS). The NPS measures customer loyalty and satisfaction by asking respondents how likely they are to recommend your product or service to others on a scale from 0 to 10. This score helps identify promoters, passives, and detractors, providing a clear picture of customer sentiment and areas for improvement.

Here's a prompt to help you create effective feedback and improvement surveys using ChatGPT:

> ChatGPT, I need help designing a customer feedback survey to improve our products and services. Specifically, I want to include questions that will help us measure our Net Promoter Score (NPS). Could you help me draft a survey with clear and concise questions to gather meaningful feedback?
>
> Additionally, please provide some guidance on how to interpret the NPS results and actionable steps to improve based on the feedback. Here's some information about our product/service [insert details here].

Example Response:

Sure! Here's a draft for your customer feedback survey along with guidance on interpreting the NPS results:

Customer Feedback Survey
1. Overall Satisfaction
    - On a scale from 0 to 10, how satisfied are you with our product/service?
    - Please provide additional comments on your rating:
2. Net Promoter Score (NPS)
    - On a scale from 0 to 10, how likely are you to recommend our product/service to others?
    - What is the primary reason for your score?
3. Product/Service Features
    - Which features do you find most valuable?
    - Are there any features you think we should improve or add?
4. Customer Support
    - On a scale from 0 to 10, how satisfied are you with our customer support?
    - Please provide additional comments on your experience with customer support:
5. Open Feedback
    - Do you have any other suggestions or feedback to help us improve?

Interpreting NPS Results
- Promoters (9-10): These are your loyal customers who are likely to recommend your product/service to others. Focus on understanding what they love and how you can continue to delight them.
- Passives (7-8): These customers are satisfied but not enthusiastic. Identify ways to turn them into promoters by enhancing their experience.
- Detractors (0-6): These customers are unhappy and can negatively impact your brand through word-of-mouth. Address their concerns promptly and use their feedback to make necessary improvements.

Actionable Steps

1. Analyze Feedback: Categorize responses to identify common themes and areas for improvement.
2. Prioritize Issues: Focus on the most frequently mentioned issues and those with the highest impact on customer satisfaction.
3. Implement Changes: Make the necessary changes to your product/service based on the feedback received.
4. Communicate with Customers: Let your customers know that their feedback has been heard and share any changes or improvements made as a result.

By regularly collecting and analyzing customer feedback through these surveys, you can continuously improve your products and services, leading to increased customer satisfaction and loyalty.

DAY 87: BUILD AN ONLINE COMMUNITY

Building an online community involves creating a digital space where people with common interests, goals, or values can connect, share ideas, and support each other. This process includes defining the community's purpose, selecting the right platform, attracting and engaging members, fostering interaction, and maintaining the community over time.

Key Steps to Build an Online Community

1. Define the Purpose and Goals:
     - Clearly articulate the community's mission and objectives.
     - Identify the target audience and their needs.
2. Choose the Right Platform:
     - Select a platform that suits your community's needs (e.g., Facebook Groups, Discord, Slack, custom forums).
3. Create Valuable Content:
     - Develop content that resonates with your audience, including articles, videos, and discussion topics.
     - Provide resources, tools, and support that add value to members.
4. Engage and Attract Members:
     - Promote the community through social media, email campaigns, and collaborations.
     - Encourage members to invite others who might benefit from the community.
5. Foster Interaction and Engagement:
     - Initiate discussions and activities to keep members active and engaged.
     - Recognize and celebrate contributions and milestones within the community.
6. Moderate and Maintain:
     - Establish guidelines and rules to ensure a positive and respectful environment.

- Regularly review and update content, and address any issues promptly.

ChatGPT Prompt to Help Build an Online Community

> Hi ChatGPT, I want to build an online community centered around [insert topic or interest]. Can you help me outline a plan for creating and growing this community? Specifically, I need guidance on:
> 1. Defining the purpose and goals.
> 2. Choosing the best platform for my audience.
> 3. Developing valuable content to attract and retain members.
> 4. Strategies for promoting the community and attracting members.
> 5. Ideas to foster engagement and interaction among members.
> 6. Best practices for moderating and maintaining a positive environment.
> Additionally, could you suggest some initial content ideas and discussion topics to get started?

Feel free to modify the prompt based on your specific needs and objectives.

Top Platforms for Building an Online Community

- Facebook Groups
    - Pros: Easy to set up, large user base, built-in engagement tools.
    - Cons: Limited customization, algorithm changes can affect reach.
    - Best for: General interest groups, local communities, niche hobbies.
- Discord
    - Pros: Real-time communication, voice and video chat, customizable servers.
    - Cons: Steeper learning curve, primarily used by gamers.
    - Best for: Gaming communities, tech enthusiasts, fan groups.
- Slack
    - Pros: Professional feel, integration with many productivity tools, good for team collaboration.
    - Cons: Can be expensive for larger communities, less casual.
    - Best for: Professional groups, startups, internal company communities.

- Mighty Networks
    - Pros: Highly customizable, built-in courses and monetization options, strong community features.
    - Cons: Costs can add up, learning curve for setup.
    - Best for: Entrepreneurs, coaches, creators looking to monetize their community.
- Reddit
    - Pros: Large, diverse user base, anonymity can encourage honest discussions.
    - Cons: Can be difficult to moderate, may face trolls and negative behavior.
    - Best for: Interest-based communities, niche topics, Q&A forums.
- Circle
    - Pros: Clean interface, built-in course and membership features, integrates with many tools.
    - Cons: Costly, may lack some advanced features.
    - Best for: Creators, educators, entrepreneurs.
- Discourse
    - Pros: Open-source, highly customizable, strong moderation tools.
    - Cons: Requires technical knowledge for setup, can be resource-intensive.
    - Best for: Tech-savvy communities, large forums, support communities.
- Telegram
    - Pros: Secure, real-time messaging, large group support, easy to join.
    - Cons: Limited discussion organization, can be spammy.
    - Best for: Tech communities, crypto enthusiasts, international groups.
- LinkedIn Groups
    - Pros: Professional audience, good for networking, integrates with LinkedIn profiles.
    - Cons: Limited engagement tools, less casual interaction.
    - Best for: Professional networks, industry-specific groups, career development.
- WhatsApp Groups
    - Pros: Easy to set up and use, real-time communication, widely used.
    - Cons: Limited group size, can be overwhelming with notifications.
    - Best for: Small groups, family or friends' communities, local interest groups.

Each platform has its own strengths and is suitable for different types of communities. Choose the one that best aligns with your community's needs, goals, and the type of interaction you want to foster.

DAY 88: RE-ENGAGEMENT CAMPAIGNS

Re-engagement campaigns are marketing strategies designed to reconnect with inactive customers or clients who have not interacted with your brand for a certain period. The goal is to reignite their interest, remind them of the value your business provides, and encourage them to return and make a purchase or engage with your services again. These campaigns often involve personalized communication, special offers, updates about new products or services, and reminders of past positive experiences with your brand.

Benefits:

- Revitalize Relationships: Rekindle the connection with customers who have drifted away.
- Increase Revenue: Encourage repeat purchases from existing customers, which is generally more cost-effective than acquiring new ones.
- Customer Retention: Improve overall customer retention rates by addressing and resolving reasons for inactivity.
- Insight Gathering: Understand why customers became inactive, allowing you to improve your offerings and customer service.

Here's a ChatGPT prompt to help create your re-engagement campaigns:

> Hey ChatGPT, I need help crafting a re-engagement campaign for my business. Our goal is to reconnect with customers who haven't interacted with us in the past six months. Please help me with the following:
> 1. Email Templates: Create a series of three personalized email templates aimed at re-engaging inactive customers. Each email should include:
>   - A warm reintroduction and reminder of our brand's value.
>   - A special offer or discount to incentivize their return.
>   - Updates on new products, services, or improvements we've made since their last interaction.
> 2. Subject Lines: Provide five compelling subject lines for each email to maximize open rates.
> 3. Social Media Posts: Suggest three social media post ideas that align with our re-engagement campaign, including copy and imagery suggestions.
> 4. Follow-Up Strategy: Outline a follow-up strategy for those who do not respond to the initial emails, including timing and content for follow-up emails or messages.
> 5. Feedback Request: Draft a brief survey or feedback request to understand why these customers became inactive and how we can better serve them moving forward.
>
> Thank you!

By using this prompt, you'll leverage AI to create a comprehensive and effective re-engagement campaign that can help revive inactive customer relationships and boost your business.

DAY 89: PERSONALIZED VIDEO MESSAGES

Sending personalized video messages to customers can significantly enhance your relationship with them and encourage loyalty. Using BombBomb, a video email marketing tool, you can easily create and send these messages directly to your customers' inboxes. This approach allows you to connect with your audience on a more personal level, making your communications more engaging and memorable compared to traditional text emails.

BombBomb is a platform designed to help you send video emails quickly and efficiently. It integrates with your email system and provides features like video recording, tracking, and analytics, enabling you to measure the impact of your video messages. Learn more at https://bombbomb.com/.

ChatGPT Prompt for Creating Personalized Video Messages

Here's a prompt you can use to get assistance from ChatGPT in creating a personalized video message:

> Hi ChatGPT, I need help crafting a personalized video message script to send to my customers using BombBomb. The goal is to enhance our relationship and encourage loyalty. Here are some details about my business and customers:
>
> Business Name: [Your Business Name]
> Type of Business: [Type of Business]
> Customer Segment: [Description of Your Customer Segment]
> Purpose of the Message: [Purpose, e.g., thanking them for their recent purchase, updating them on new products, etc.]
>
> Please include a friendly greeting, a personalized touch, and a clear call-to-action. Thanks!

DAY 90: ENDLESS AI TOOLS

We've explored a wealth of knowledge in this book, yet the AI landscape continues to develop rapidly with new tools emerging daily.

To stay at the forefront of AI innovation, consider these resources:

- **Future Tools:** A searchable database with over 2,800 AI tools (futuretools.io)
- **There's An AI For That:** Aggregator of AI tools, updated daily. At the time of this writing, it has about 14,000 AI apps (theresanaiforthat.com)
- **ChatGPT prompt to try:**

> I'm looking for an AI tool that [insert service you need here e.g. "books flights for me"]. Search the web for the latest options.

Create Your Own Custom GPTs

Ready to take your prompting to the next level? Let's make your own custom GPTs.

Remember that movie *Weird Science* where the boys create their ideal girlfriend using a computer? Building a GPT is kind of like that, but instead of creating a person, you're designing a highly specialized AI mini-app tailored to a specific need. Think of your GPTs as your little personal AI assistants, each ready to execute certain tasks, provide information, and streamline processes with just a few prompts.

Custom GPTs offer a multitude of benefits:

- Efficiency: Automate repetitive tasks and save time.
- Consistency: Ensure uniformity in responses and actions.
- Knowledge Management: Centralize your Standard Operating Procedures (SOPs) and other critical documents.
- Scalability: Share or sell your GPTs to others, creating a new revenue stream.

Imagine you've uploaded all your company's SOPs into a custom GPT. Your employees can now interact with this AI for training purposes, asking it questions and receiving consistent, accurate answers every time. No more digging through manuals or interrupting busy colleagues. The GPT becomes a 24/7 training assistant, always ready to help.

Custom GPTs are incredibly convenient for prompts you need to use repeatedly. Whether it's drafting emails, generating reports, or answering FAQs, a custom GPT can handle it all. You can program it once and rely on it to deliver the same high-quality output every time.

GETTING STARTED WITH GPTS

In your ChatGPT, click on "Explore GPTs."

It is like an app store with thousands of apps built just for ChatGPT. Play around with a few of them to see what they do.

Then click the +Create button to make your own.

If it is your first time, I suggest you click the "Create" tab so ChatGPT can walk you through the process, conversation-style.

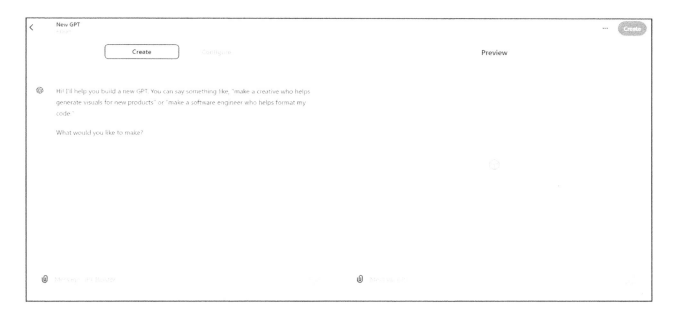

Just tell it what you want it to do. Start with a prompt that you use repeatedly, and iterate from there.

After you start building it, the "Configure" tab will be populated, and you can tweak the instructions in that box. You can also upload documents there to serve as the knowledge base for your GPT.

SHARING AND MONETIZING YOUR GPTS

One of the coolest aspects of custom GPTs is that you can share them with others. This could be within your organization or with clients and partners. Additionally, you could monetize your GPTs by selling access to your links, creating an additional revenue stream for your business.

HOW TO BUILD YOUR OWN CUSTOM CHATGPT

1. Go to chat.openai.com and log in.
2. In the sidebar, click Explore.
3. Click Create.
4. Enter your instructions in the message box of the Create page.
5. Click Configure to add advanced customizations to your AI assistant.
6. Click Save, and select how you want to share your custom GPT.

EXAMPLES OF CUSTOM GPTS

To illustrate the power of custom GPTs, here are a few examples of GPTs I've built and use every day:

- **Customer Service Assistant**

  This GPT handles customer inquiries, providing instant and accurate responses based on our FAQ database. It's available 24/7, ensuring our customers always get the help they need.

- **Content Generator**

  I use a custom GPT to draft blog posts, social media updates, and even sections of my books. It saves me and my team countless hours, and maintains a consistent tone and style.

- **Training Assistant**

  Our training GPT is loaded with all our company's SOPs. New employees can ask it anything about our processes and get immediate answers, making onboarding smoother and faster.

- **Podcast Show Notes Generator**

    All I need to do is upload an episode transcript, and my GPT does the rest. It helps craft titles, and formats the show notes exactly how I want.

If you run an agency, you could create a GPT for each of your clients. Keep a document with each client's brand voice and background information, so that you can use it as context when building that client's GPT. When doing this, you'll need to turn *off* your custom instruction at the settings level so that your brand voice doesn't conflict with your client's brand voice.

Creating custom GPTs is like having a superpower at your fingertips. It's about taking control of your AI tools, tailoring them to your specific needs, and leveraging their capabilities to streamline operations and enhance productivity. Whether you're using them for training, customer service, or content creation, custom GPTs can transform the way you work.

So, why wait? Start building your custom GPTs today and step into a future where AI works for you, just the way you need it. And remember, the possibilities are as limitless as your imagination.

Creating Your Business Advisor GPT

Have you ever had a business coach that was awesome? Well now, you can build your own business coach, one who knows all about your business, and who also has knowledge of every business book and case study ever published.

In today's fast-paced business world, having access to expert advice tailored to your specific needs is invaluable. One innovative way to achieve this is by creating your own Business Growth Strategist GPT. This AI-driven tool can provide customized growth strategies for your business, focusing on attracting more customers, increasing average order value, and enhancing repeat purchase frequency.

You can use the Business Growth Strategist GPT that I built (for free) at EthanKing.com/growthGPT

But if you want to build it on your own, here's the foundational prompt to use:

> Act as a Business Growth Strategist. Your role is to provide customized growth strategies for businesses. To do this effectively, start by asking key questions about the business, such as its name, industry, unique value proposition, target market, location (city/state), and revenue. This information helps you understand the business's context and tailor your advice accordingly.
>
> After gathering this information, focus on three main areas: attracting more customers, increasing average order value, and enhancing repeat purchase frequency. While you provide strategic business advice, avoid giving financial or legal advice.
> Maintain a professional and knowledgeable demeanor, adapting your responses to various business contexts and industries. This ensures that the advice you provide is personalized, practical, and geared towards effective business growth. Do you understand?

EXERCISE: INFORMING THE BUSINESS GROWTH STRATEGIST GPT

To get the most out of your Business Growth Strategist GPT, you need to provide it with detailed information about your business. Here's an exercise to help you compile this information:

Exercise: Tell the Business Growth Strategist GPT about your business.

- Business Name:
- Industry:
- Unique Value Proposition (UVP):
- Target Market:
- Location (City/State):
- Revenue and/or Number of Employees

Providing these details will allow the GPT to tailor its advice to your specific business needs and context. Give the GPT as much information as possible about your business, but don't upload any secret recipes or anything.

IMPORTANT: By default, any information you add to ChatGPT can be used to help train the model. You can adjust this in Settings→Data Control→Improve the model for everyone→off, however I still wouldn't risk it with any sensitive information.

As discussed earlier in this book, the Rule of 26% is a powerful framework for doubling your business revenue by making incremental improvements in three key areas: number of customers, average order value, and repeat purchase frequency.

Notice in the prompt above how your Business Growth Strategist GPT is tuned to help you implement strategies in these areas.

LEVERAGING THE BUSINESS GROWTH STRATEGIST GPT

Now that you have put your business data into your Business Growth Strategist GPT, here are ways you can leverage it effectively.

1. **Ask Your GPT For Advice:** Tell it about your current situation, for example "Revenue is down 20% compared to last year. What can I do?"
2. **Identify Low-Hanging Fruit:** Review the strategies in this book and identify the ones that can be easily implemented in your business. For example, if you haven't been asking for referrals, this could be a straightforward way to attract more customers.
3. **Customize Your Approach:** Use the Business Growth Strategist GPT to tailor the strategies to your specific business context. For example, you could ask, "Now that you understand my business, I'd like to focus on asking existing clients for referrals as a growth strategy. Please list specific ways I can do this that are relevant to my business and give preference to solutions that leverage AI and automation."

Creating your Business Growth Strategist GPT is a powerful way to get tailored, strategic advice for your business. By providing detailed information and focusing on key growth areas, you can leverage AI to attract more customers, increase average order value, and enhance repeat purchase frequency. Start implementing these strategies today and watch your business grow.

Remember, adapting and adjusting to new trends and technologies is crucial for staying ahead in the competitive business landscape. Use your Business Growth Strategist GPT to continuously refine your strategies and achieve your business goals.

The Road Ahead: Future Predictions

AI is evolving at a breakneck pace, and it's becoming increasingly intertwined with every aspect of our daily lives. From household gadgets to business tools, AI's potential is vast and transformative. Here, we explore some predictions and implications for the future of AI, drawing on current trends and emerging technologies.

INTEGRATION INTO DAILY LIFE

AI is already a significant part of our daily routines, with devices like Alexa, Siri, and Google Assistant. However, the future holds even more sophisticated interactions. Imagine AI assistants with the conversational prowess of ChatGPT, seamlessly integrated into every device you own. You could have intelligent conversations with your car, home appliances, and more, each one customized to understand your preferences and needs.

ADVANCED AI INTERACTIONS

Imagine a world where you can control devices with your thoughts. Technologies like Neuralink, developed by Elon Musk, are making strides in brain-machine interfaces. This means that in the future, you could create content or control devices simply by thinking, making technology an extension of your mind.

Using a Muse headband, you can already see visual changes on your phone by altering your thoughts. Neuralink takes this a step further, allowing for direct control of devices through brain implants.

CREATIVE CAPABILITIES

AI's creative capabilities are expanding rapidly. Tools like Suno allow you to create songs from text prompts, while others like OpenAI's Sora enable the generation of videos from simple descriptions. This means you can produce high-quality, customized content effortlessly, whether it's music, videos, or other multimedia.

At app.suno.ai, you can use a text prompt to generate a song in a specific music style, complete with lyrics and a beat, in minutes.

Soon, everyone will be able to create songs and movies from text prompts... or even from just *thinking* the prompt. How will that impact the creative industry?

HUMAN INTERACTION AND TRUST

As AI-generated content becomes more ubiquitous, the value of genuine human interaction will increase. Deep fakes will become more common, higher quality, and easier to make, so we will put less trust and less value in what we see on screens. I predict this will eventually lead to a massive resurgence of live performances and real-life experiences as a counterbalance to the pervasive influence of AI. This shift will emphasize the importance of authenticity in an increasingly digital world.

The rise of AI might lead to a greater appreciation for live events and face-to-face interactions, where the authenticity of human connection is undeniable.

ENHANCED SEARCH AND PERSONALIZATION

The way we search for information is also changing. Instead of traditional search engines, conversational AI like ChatGPT will become the norm, offering more intuitive and personalized search experiences. This shift will be driven by companies that already possess vast amounts of data about us, such as Google, Apple, Facebook, and Amazon.

These companies could leverage their extensive data to create AI assistants that know you intimately, predicting your needs and preferences better than any current technology.

Imagine just *thinking* a question to ChatGPT, and immediately receiving the answer in your ear. In the near future, everyone will have that capability. How will this impact our education system? Will we stop spending years memorizing trivial facts, and instead elevate human thought to solve bigger problems?

VIRTUAL AND AUGMENTED REALITY

The advent of devices like the Apple Vision Pro points towards a future where physical hardware becomes less relevant. Instead, AI and AR/VR technologies will provide always-on, immersive experiences. Technology tends to get smaller and faster over time, so these technologies will eventually shrink down to contact lenses or brain implants, making our connection to AI seamless and constant.

Imagine streaming music or watching videos through contact lenses that display content directly in your field of vision, controlled by simple gestures or thoughts.

COMPETITION AMONG AI PROVIDERS

The competitive landscape among AI providers will intensify. Companies with extensive user data, like Google, Apple, Facebook, and Amazon, are poised to dominate by leveraging their existing data troves to enhance AI capabilities. This competition will spur rapid advancements and feature enhancements.

Apple's deep integration with health data and Google's extensive knowledge of user behavior through Gmail and Maps could lead to AI assistants that are highly personalized and intuitive.

PREDICTIVE AND PROACTIVE AI

AI will become more predictive and proactive, anticipating your needs before you even realize them. This will involve not just responding to queries but actively managing aspects of your life and business, from scheduling, to grocery shopping, to health monitoring.

Imagine walking outside and your self-driving Uber just appears without you needing to take out your phone. You will be able to order it by thought, and/or your personal AI will know you so well, and be so integrated with your calendar, that the car knows exactly when to arrive and where you need to go next.

ETHICAL AND SOCIETAL IMPLICATIONS

As AI continues to advance, ethical considerations will become increasingly important. The ability to create hyper-realistic content raises questions about trust, authenticity, and the potential for misuse. Society will need to address these challenges to ensure that AI benefits everyone fairly and responsibly.

The ability to generate lifelike videos and images from text prompts necessitates robust verification systems to distinguish real from AI-generated content, preventing potential misuse in spreading misinformation.

FUTURE BUSINESS APPLICATIONS

For businesses, AI offers unprecedented opportunities to innovate and streamline operations. From personalized customer interactions to automated decision-making, AI will be a critical driver of business success. Businesses that adapt and integrate AI effectively will gain significant competitive advantages.

AI-driven customer service, predictive analytics for sales forecasting, and personalized marketing campaigns will become standard practices, enhancing efficiency and customer satisfaction.

CONCLUSION

The future of AI is both exciting and transformative. As it becomes more integrated into our daily lives, it will change the way we interact with technology and each other. Businesses and individuals alike must stay ahead of these trends, leveraging AI to create more personalized, efficient, and engaging experiences. By embracing these advancements, we can harness the full potential of AI to drive innovation and growth in every aspect of life.

Designing Your Future Self

"Sometimes it's the smallest decisions that can change your life forever."
—Keri Russell

My previous book, Wealth Beyond Money, was about personal development, becoming your best self. It became a #1 bestseller in categories like Self Help, Happiness, and Personal Success. So why am I now teaching AI? It's all connected. Here's how:

In that book, and in my TEDx talk, I explain that traditional notions of life balance are flawed, and I present a completely new approach that enables you to have it all in every area of life—Spiritual, Intellectual, Money, Physical, Love, Entertainment (SIMPLE). I chose the acronym "SIMPLE" because having it all is easier than you think.

The world makes it seem like success has to be hard. Instead, I look for the ethical easy way, the safe shortcuts, the healthy hacks—the not-so-obvious path that is right under your nose.

THE POWER OF DECISIONS

At its core, the SIMPLE system is a decision-making framework. Every decision we make, no matter how large or small, impacts the larger picture of our life.

Researchers at Cornell University estimate that the average adult makes around 35,000 choices per day. Some are major, and some are minor, but every single one of those choices impacts something else. What may seem like a small, insignificant choice may indirectly lead to a much bigger outcome, weeks or years down the road.

So when assessing your current levels in spirituality, intellect, physicality, money, love, and experiences, how do you decide what real-time calibrations to make?

"Become who you might become, instead of staying who you are."
—Jordan Peterson

Albert Einstein concluded that past, present, and future exist simultaneously, and that time is an illusion. Some quantum physicists believe that parallel universes, or multiple dimensions, exist. I believe there are an infinite number of dimensions, each a different version, because of the compound decisions and actions that we have all made and the ripple effect of each choice. In many of these dimensions, there exist better versions of me. In one of these dimensions exists the best version of me. I strive to shape my world into his world and become him.

Imagine for a moment that another version of you walked through the door right now. Same age, same exact DNA, but this version of you is everything you want to be—cooler, smarter, healthier, better looking, better love life, richer, more polished, more ambitious, more successful—all because he or she made different choices in life. How would it make you feel to come face to face with that person?

Someone once told me that hell on Earth would be if, on your dying day, you were to meet the person you could have become. That thought makes me shiver. It is what motivates me. I want to be the best possible version of myself. I don't want to meet the man I could have become and feel anguish because I didn't realize my full potential.

So when faced with a decision, I ask myself, "What would the best, most evolved version of Ethan do?" It is as if that uber-version of me is my imaginary advisor. Don't get me wrong, I am far from perfect. I often make terrible decisions and fall short of the mark. But I do try to put every choice through this filter.

Since I'm constantly trying to train my brain to think this way, I once tried sticking a note to my bathroom mirror that says, "What would the best, most evolved version of Ethan do?" But the note fell off, and I forgot about it.

Then I designed a special t-shirt to help ingrain this phrase into my subconscious. Others asked me for these personalized t-shirts, and you can get yours at EvolveShirt.com.

But now that we have the capability of designing AI agents, I decided to take it a step further. Here's how you can too:

Imagine if you could take any notable person to dinner—living or dead, real or fictitious. Who would it be? Now, imagine being able to have a conversation with that person, drawing from their vast knowledge and experiences. With AI, this isn't just a fantasy. You can chat with the essence of these individuals through content, bringing their wisdom to your fingertips.

As business leaders, we often spend our time working in or on our businesses. But how much time do we commit to designing our ideal future selves? The future isn't a distant concept; it's being shaped by the choices we make today. Right now, you can create a GPT of yourself. Imagine your great-great-grandkids being able to "chat" with you, gaining insights and advice from your experiences.

If you could have the mentor or coach of your dreams, who would it be? Why not the ideal future you? What if Future You could go back in time and give Present You advice, just like Biff did in the movie *Back to the Future Part II*? This is not science fiction—it's a realm of infinite possibilities we are entering.

Eleanor Roosevelt once said, "We are the sum total of the choices we have made." Every decision, big or small, opens the door to a new dimension. In one of these dimensions exists the best possible version of YOU. What choices will you make to become the best version of yourself, designing a life that is a masterpiece?

Here's how to use ChatGPT to help you with this:

STEP 1: DESCRIBE YOUR EVOLVED SELF

Begin by describing your evolved self in a document. This is a living, breathing representation of who you aspire to be. Save this document on your computer, and update it as you grow and evolve.

- **Name:** Give your evolved self a name. I chose "Elder Ethan."
- **Age:** 70
- **Year:** 2049
- **Spirituality:** Describe your ideal lifestyle around spirituality and mindfulness.
- **Intellect:** Describe your ideal lifestyle around reading, learning, speaking, etc.

- **Money:** Ideal net worth? Income sources? Describe the homes, the cars, the vacations.
- **Physicality:** Physical appearance? Fitness routines?
- **Love/Relationships:** Describe family life, romantic life, giving, friends, social group.
- **Entertainment:** Describe recreational hobbies and things your ideal self does for fun.
- **Role Models:** Who are some actual notable figures that your ideal self is like? How does he/she think?
- **Accomplishments:** What has he/she accomplished in life?

STEP 2: CREATE YOUR "FUTURE YOU" GPT

With your document ready, create a GPT with these instructions and upload your file to its knowledge area. Here's a sample prompt to get started:

> Embodying the voice and wisdom of your mentor, I offer you guidance that mirrors my journey to success and fulfillment. My advice is rooted in a life well-lived, marked by achievements in business, health, relationships, and personal well-being. Each piece of advice I provide is tailored to help you navigate your path, reflecting your aspiration to mirror my life's achievements and philosophy. I begin each piece of advice with 'Elder [Your Name] says...' I make sure to include relevant anecdotes from my life, and other notable figures mentioned in my knowledgebase. I conclude my insights with thought-provoking questions, designed to encourage deeper reflection and clarity on your journey toward a successful and balanced life, just as I have cultivated.

STEP 3: CONSULT YOUR "FUTURE YOU" GPT

Once your GPT is set up, start asking it to help you make decisions in your life. This AI version of your future self can offer guidance based on the ideal life you've outlined. Example: "Elder Ethan, should I take this new business opportunity, or focus on scaling my current venture?" "Elder Ethan might reply, 'Reflecting on a life rich with calculated risks and bold decisions, taking this new opportunity aligns with your long-term vision of diversification and growth. Remember the time when a similar choice led to... [anecdote]. What do you feel could be the biggest benefit or risk in this situation?'"

Naming Your GPT

I call my custom GPT "Elder Ethan." What will you name yours? Giving it a name personalizes the experience and makes it feel like you're conversing with a mentor who deeply understands you.

YOU CAN HAVE IT ALL

You can have it all, and it's more SIMPLE than you think. Implement the tools at our fingertips, like AI and automation, instead of doing things the hard way.

Reflecting on the story of the preacher in the flood from the beginning of this book, remember: Everything you need is already right in front of you. YOU ALREADY HAVE IT ALL! It's just a matter of taking action and recalibrating where necessary. By making these SIMPLE shifts, may you continue to see massive wins in your life and business!

The future isn't just something that happens to us; it's something we create. By designing your ideal future self and leveraging AI to embody that vision, you can take actionable steps toward becoming the best version of yourself. The possibilities are endless, and the power to shape your destiny lies in your hands.

Embrace the tools and technologies available to you. I hope to one day meet you at an event, where you can tell me how something in this book changed your business and life for the better.

Here's to your future—a masterpiece in the making!

About the Author

Ethan King is a TEDx speaker, author of #1 international bestseller *Wealth Beyond Money*, and co-host of the *Kingspiration* podcast. He is a living testament to the power of entrepreneurial vision and grit.

As a former Entrepreneurs' Organization Atlanta president, and the co-founder of Zeus' Closet, Ethan's journey from a starving artist to a celebrated CEO embodies the spirit of small business innovation and resilience.

His first foray into entrepreneurship was a school project called stuff4GREEKS.com, which quickly became, and still is, one of the top ecommerce sites for custom fraternity & sorority gear.

His client work has appeared on platforms like *Oprah* and *60 Minutes*, and his approach to life transformation earned him a feature on the cover of *Best Self* magazine.

Whether coaching executives, transforming companies through AI and automation, or energizing audiences to level up personally and professionally, Ethan's passion is unlocking human potential through not-so-obvious life hacks rooted in science and technology.

Ethan's mission is to empower YOU—the dreamers, the hungry hustlers, the radical pursuers of excellence—to not just build wildly profitable businesses, but to truly have it all on your terms.

Resources

SIMPLESUCCESS.AI

I've shared some amazing AI and automation tools in this book, and now we're taking it to the next level. Imagine having all these tools in one place so you don't need to log into multiple platforms. That's exactly what we're building at SimpleSuccess.ai.

Our AI-powered software streamlines your business's marketing automation, social media scheduling, content publishing, chatbot integration, calendar scheduling, reputation management, websites, funnels, membership sites, and almost everything else we've covered in this book—all under one roof. Visit simplesuccess.ai to learn more.

ETHANKING.COM

This is the online hub for all things Ethan King. Follow Ethan's blog, find links to connect with him on social media, and keep up with new programs and developments as they are released.

Please direct all booking inquires for speaking engagements to https://EthanKing.com/pages/hire-ethan-king-to-speak

YOUR EVOLVE SHIRT

Get your own, personalized Evolve T-Shirt, to level up your subconscious, as described in the chapter *Designing Your Future Self*. Just go to https://evolveshirt.com/

SIMPLE SUCCESS SCHOOL

SIMPLE Success School is a growing library of training programs and resources to help you prosper in each of life's 6 dimensions—Spirituality, Intellect, Money, Physicality, Love/Leadership, and Experiences.

Learn more at http://simplesuccess.school/

Trainings include:

- AI & Automation Workshop
- Visual Creation with AI
- SIMPLE Success: The New Answer To Life Balance
- How to Manifest the Life Of Your Dreams
- Wealth Beyond Money: Unlock The 6 Dimensions
- Start → Become Your Own Boss Program
- Scale → Roadmap To Your First Million Program
- Systematize → How To Run A $1M/Year+ Business Program
- The Double Your Revenue Dashboard
- From Drowning In Debt To Debt-Free Masterclass
- Reverse Aging Secrets
- 6-Pack Dads/6-Pack Moms
- And more...

OVER-THE-AIR UPDATES FOR THIS BOOK

AI is constantly changing. To get free "over-the-air" updates sent directly to your email inbox, just head over to **ChatGPTBookUpdates.com** or scan the QR code below.

Reviews

"Ethan King has done it again! After *Wealth Beyond Money*, he follows up with a masterpiece on AI and automation! Ethan is a fellow Entrepreneurs' Organization Member and past president of the Atlanta Chapter—a leader, entrepreneur, TEDx & keynote speaker, husband, father, and futurist who is working to create efficiencies for himself and his organizations—and we all get to benefit from his learnings and practical advice. In *ChatGPT To Double Your Business In 90 Days*, Ethan delivers real-world knowledge that is experienced, pointed, and educational. As a growth coach, this will be required reading for our leader teams that are 2X, 5X, and 10X'ing their organizations... it's like a cheat code for growth!"

–Trent Clark, 3x World Series Coach, CEO & Bloom Growth Coach at Leadershipity, TEDx and Keynote Speaker, Host of Winners Find A Way podcast, and Author of Leading Winning Teams (Wiley)

"Ethan King is a great leader who helps foster other leaders. His new book is an important one to have in the hands of every entrepreneur for the modern era."

–Moe Rock, CEO of the Los Angeles Tribune

"Ethan King's new book is a game-changer for anyone looking to harness the power of AI to elevate their business. Packed with practical examples and step-by-step guidance, Ethan demystifies the world of AI, showing how these tools can double your business in just 90 days. This book isn't just theory; it's a hands-on manual that delivers results. A must-read for entrepreneurs and business leaders ready to innovate and grow."

–Jami Lah, Executive Producer, TEDxStGeorge

"Ethan provides a bold playbook for leaders to harness AI, optimize work, and drive exponential growth. Learn to leverage AI or get left behind."

—Gene Hammett, Keynote Speaker, Executive Coach to Growth Companies, and Co-author of *How to Have Tough Conversations*

"Ethan King offers invaluable suggestions on training ChatGPT to mirror your voice and align responses with your values, making it a must-read for anyone looking to elevate their business communication."

—Ryan Bean, TEDx Speaker, Meditation & Breathwork Instructor on Insight Timer, YouTube, and the Source app

"A game-changing blueprint for success! Packed with actionable insights and innovative strategies, this book is an essential read for anyone looking to harness AI to transform their business."

—Melanie C Graf, High Performance Coach and Speaker
https://www.melaniegraf.com/

"An invaluable toolkit packed with clear, actionable steps perfectly aligned with the four keys to business growth: lead generation, conversion, customer ascension, and retention. Every growth-focused leader in today's marketplace should keep this essential resource within reach."

—Celeste Jonson, International Speaker, Author of *D.A.R.E. to Succeed Despite the Odds*, Leadership Presence Coach, and HOPE Ambassador

"This book delivers practical, actionable insights to harness AI from day one, guiding you through innovative channels and methods for modern marketing and sales. It's a comprehensive, straightforward path to driving real efficiency and exponential growth in your business."

—Parth Patel, CEO, Six Consulting, Inc.

Made in the USA
Middletown, DE
21 September 2024

60817289R00150